Data Analytics in the Era of the Industrial Internet of Things

Aldo Dagnino

Data Analytics in the Era of the Industrial Internet of Things

 Springer

Aldo Dagnino
ABB Group
Cary, NC, USA

ISBN 978-3-030-63141-3   ISBN 978-3-030-63139-0   (eBook)
https://doi.org/10.1007/978-3-030-63139-0

This Springer imprint is published by the registered company Springer Nature Switzerland AG
The registered company address is: Gewerbestrasse 11, 6330 Cham, Switzerland

To my wife Carla for her support, patience, and love and to my parents for the important and valuable lifelong lessons they taught me – Aldo Dagnino

Preface

This book has several objectives aimed at increasing the understanding of systems developed to support the Industrial Internet of Things (IIoT). It is important to help the reader understand the IIoT phenomenon as it is an industrial approach that is and will be disruptive in organizations such as manufacturing, service, process industries, and retail, among others. The underlying principles of an IIoT organization are presented and also a discussion on how new technologies such as sensors, wireless, computing, data storage, artificial intelligence and machine learning, high-performance computing, and leading-edge visualization enable the IIoT. The book also provides the reader with a description of a typical IIoT architecture that industries implement to efficiently operate. An important objective is also to show how the IIoT brings about a deep change in business models of providers and consumers of products and services and present examples on the transition between current business models and IIoT business models. One of the important elements of the book is to provide the reader with basic understanding of the power of data analytics, which is the primary engine of IIoT systems. A simple-to-understand discussion on different machine learning models is presented so that business readers, without computer science or engineering background, can follow how these machine learning models operate on data. The readers are also exposed to real-world industrial cases where advanced analytics and machine learning have been used to achieve a business goal, address a business pain point, or crystalize a business opportunity. An important element that is underlined in this book is that technology is not in all cases a vehicle to replace human workers or domain experts. I have been fortunate to work with organizations that are very progressive in terms of technology utilization but also are conscious that these new technologies should be implemented with the idea not only to improve the business but also to empower subject matter experts and workers and make their work more productive, less boring, and more exciting, as well as facilitate and enrich their decision-making capabilities.

The targeted audience of this book includes managers, business professionals, engineers, and students that wish to learn how analytics is used to address business opportunities in the IIoT and bring organizations closer to a new and more productive business model. It is my hope that this book will spark interest in the audience to further understand how data can be used as powerful raw material to extract knowledge jewels within the IIoT framework to improve business outcomes, reduce waste of raw materials and resources, increase level of workers' knowledge, increase employee satisfaction, enhance creativity, and help employees to design solutions that are creative and use new technologies.

The Industrial Internet of Things, from my perspective, is a phenomenon based on an interdisciplinary point of view of the modern world. The IIoT is the result of several elements meeting each other in a contemporaneous way. First, recent advances of sensor technologies have been powered by high-speed and low-cost electronic circuits, innovative signal processing methods, and novel advances in manufacturing technologies. Ground-breaking sensor structures have been designed that allow self-monitoring and self-calibration of sensors. The rapid progress of sensor manufacturing technologies has allowed the production of systems and components with a low cost-to-performance ratio. Moreover, miniaturization fosters the deployment of sensors in many areas of industrial systems. Second, the development of high-performance computing allows for deployment of complex machine learning models that can analyze high volumes of sensor data very fast. Third, the phenomenal increase of data storage capabilities allows industries to save an almost infinite amount of structured and unstructured data in data warehouses. Fourth, the advent of powerful visualization tools provides the ability to deploy innovative ways for users to interact with computer systems. Fifth, the advent of cloud computing provides a capability to perform calculations at the cloud but also at the edge or equipment level. Sixth, the networking capabilities available allow high-performance connectivity among IIoT components.

The structure of the book presents a systematic discussion of the topic in its various chapters. After the Introduction chapter, Chapter 1 is dedicated to describing the concept of the Industrial Internet of Things. A detailed discussion on the general architectural framework of IIoT is presented to understand how to perform edge-based and cloud-based analytics. A layered view of the major IIoT components are described in detail. A description of how data analytics in general and machine learning in particular play the role of the main engine in the IIoT is presented. Chapter 1 also discusses how IIoT in general and machine learning in particular are technologies that enhance the quality of work of knowledge workers and promote hiring new highly talented subject matter experts. Finally, Chapter 1 provides examples of new business models generated by the IIoT, and in my opinion they will thrive in the future.

Chapter 2 provides a description of machine learning and some important modelling specially tailored for managers and non-experts in the field. This chapter offers a classification of machine learning models. A description of each machine learning model used in the examples of the book is presented in a non-expert fashion so businesspeople can understand how the models operate on the data they analyze. This part of the book is a fascinating trip into better understanding the "magic" that AI/ML can deliver when analyzing data, but it also shows that such "magic" can be very fragile as the quality of data, the assumptions required, and the interpretation of results dramatically influence the final results and how they are used in the business world.

The following chapters in the book provide real-world examples from industry on how analytics is applied to address specific business problems or opportunities. Chapter 3 examines how ML can address the problem of finding faults in power distribution grids. Chapter 4 presents an analytics approach to analyze events and alarms in control systems. Chapter 5 discusses a remote condition monitoring analytic approach to conduct sensor data analysis and identify anomalies in rotating machines. Chapter 6 reviews an interesting machine learning approach to identify new opportunities of sales for services and products in an IIoT organization. Finally, Chapter 7 is dedicated to discussing approaches to better manage advanced analytic projects and improve the probability of their success in the IIoT organization.

Cary, NC, USA   Aldo Dagnino

Acknowledgments

The analytic applications described in this book are the result of many projects where the author collaborated with several colleagues and friends. Their contributions in the projects were invaluable and I would like to thank their dedication and professionalism. With the peril of not mentioning all of them, I would like to thank Karen Smiley, Eric Harper, Mithun Acharya, Alok Kucheria, Marcel Dix, Martin Hollander, Benjamin Kloepper, Jinendra Gugaliya, Martin Naedele, and Roland Weiss, among others. I would also like to thank my neighbor Kevin Johnson who helped me with yard work so I could meet the writing deadlines of this book.

Contents

Abbreviations

AI	Artificial Intelligence
CMMI	Capability Maturity Model Integration
IoT	Internet of Things
IIoT	Industrial Internet of Things
ML	Machine Learning
OEM	Original Equipment Manufacturer
RMP	Revolutions per Minute
SMEs	Subject Matter Experts
ERP	Enterprise Resource Planning
NPS	Net Promoter Score
MVP	Minimum Viable Product
UX	User Experience
SEI	Software Engineering Institute
SPI	Software Process Improvement
IOU	Investor Owned Utility
IED	Intelligent Electronic Devices
NN	Neural Networks
KSVM	Kernel Support Vector Machines
DCS	Distributed Control System
SAM	Smart Alarm Management

Chapter 1
Industrial Internet of Things Framework

When reading the literature available, the reader will realize that there are many definitions and views regarding what the Internet of Things (IoT) and the Industrial Internet of Things (IIoT) are. One relevant definition that is used as a fundamental concept in this book of the IoT is: "[a]n open and comprehensive network of intelligent objects that have the capacity to auto-organize, share information, data and resources, reacting and acting in face of situations and changes in the environment" [9, page 165]. Another relevant definition of the Internet of Things (IoT) is given as "people connected to people they care about through everything being connected" [10, page 5]. Gilchrist [5] discusses the power of 1% as it refers to the IIoT. Based on companies that implement an IIoT platform, it is only necessary to reduce operational costs or to reduce inefficiencies by 1% in an organization to more than recover the costs associated with implementing an IIoT platform.

The Industrial Internet of Things (IIoT) is a special manifestation of the IoT in the industrial world and allows instrumented objects (equipment, products, systems, etc.) and people to collaborate, share information, work together, and make decisions in industrial environments [6]. "The Industrial Internet provides a way to get better visibility and insight into the company's operations and assets through integration of machine sensors, middleware, software, and back end cloud compute and storage systems. Therefore, it provides a method of transforming business operational processes by using as feedback the results gained from interrogating large datasets through advanced analytics." [5, page 3]. The IIoT facilitates hundreds of smart interconnected objects and people to exchange information, adapt to their environment, continuously communicate with each other and work collaboratively to enhance production environments. Chou [2] developed a general Internet of Things framework for smart objects, which is extended in this book to include people in industrial setups. The IIoT framework presented can be viewed as the amalgamation of data generated by machine sensors and human sensors deployed at all levels in the IIoT. This extended framework is called the IIoT Framework and is presented in Fig. 1.1.

© Springer Nature Switzerland AG 2021
A. Dagnino, *Data Analytics in the Era of the Industrial Internet of Things*,
https://doi.org/10.1007/978-3-030-63139-0_1

Fig. 1.1 Industrial internet
of things framework

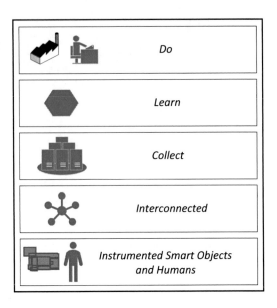

The first layer at the bottom of Fig. 1.1 shows the main actors in the IIoT, which include *Instrumented Smart Objects* and *Humans*. Instrumented smart objects can be equipment, products, things, and physical systems that have a variety of sensors that collect data from the environment where they operate. Instrumented objects can exhibit "intelligent" behavior, which means these objects use the data and information they gather from their operating environment and adapt their behavior and responses accordingly. Some intelligent actions of instrumented smart objects can be viewed as more intelligent than other actions. Inherently, the concept of intelligence is not only related with a specific task being performed but also with the passage of time. What was considered machine intelligence ten years ago, nowadays is a "routine" behavior and not particularly "high intelligent". The concept of intelligence in machines evolves as technology becomes more powerful. The second element in this bottom layer is *Humans* who work in the IIoT ecosystem and interact with each other, the instrumented objects to carry out their productive tasks, and with the external world (such as customers, suppliers, competitors, etc.). Humans in the IIoT environment include personnel in the IIoT organization (factory, process plant, etc.) that work to deliver products or services and external stakeholders that interact with the personnel (such as suppliers, customers, etc.). Humans in the IIoT also generate data as they interact with each other and with the intelligent objects. Human data is often generated in the form of unstructured data, such as the one contained in a variety of documents exchanged during the production process or the product and service delivery. For the purposes of our future discussion, the IIoT environment can be viewed as *IIoT Entities* (machines, personnel, and external stakeholders) that are involved with the production activity of the IIoT facility that generates a variety of data (structured and unstructured) emanating from either *Machine Sensors* or *Human Sensors*.

IIoT entities are *Interconnected* via the Internet or the organization Intranet as represented in the second layer from the bottom in Fig. 1.1. This interconnectedness allows instrumented smart objects and humans to exchange data, information, and messages in a secure and high performing way using appropriate communication protocols. The development of secure, high performing, and scalable wireless technologies provide more flexibility for IIoT entities to communicate within the system and with each other. Humans also exchange a variety of data via the network among each other and instrumented smart objects. Figure 1.2 shows how data and communication can flow among instrumented smart objects and humans in the IIoT. Data and communications can flow among instrumented smart objects (machines, products, etc.), and this data is primarily structured data (such as temperature thresholds, vibration thresholds, parts processes in a machine on the shop floor, messages about machine health, etc.). Connectivity involves three types, one-to-one, one-to-many, many-to-many [12].

One-to-one connection is when an individual object, product, or entity connects to a user, manufacturer, or another object. For example, a turbine that sends a message to a Service Engineer informing that there is an anomaly in the recorded vibrations in a bearing in the system. One-to-many connection occurs when a system is connected to many objects continuously or intermittently. For example, in the case where several power transformers in utility are connected to a common asset health monitoring system to identify if one of the transformers has a higher than normal concentration of gasses dissolved in their coolant oil. Many-to-many connectivity involves the situation where multiple objects to many other products or humans. As is the case of robots in an automotive manufacturing company that assemble metal parts and then they move to another robotic cell where welding is performed on the assembled parts. Interconnection capability also allows information to be exchanged

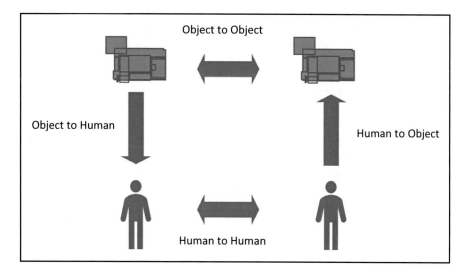

Fig. 1.2 Data exchange among IIoT entities

between the product and its operating environment and enables some functions of the object to exist outside of the actual device in what is known as the object cloud.

Instrumented smart products are emerging across many manufacturing sectors. For example, [i]"n heavy machinery, Schindler's PORT Technology reduces elevator wait times by as much as 50% by predicting elevator demand patterns, calculating the fastest time to destination, and assigning the appropriate elevator to move passengers quickly. In the energy sector, ABB's smart grid technology enables utilities to analyze huge amounts of real-time data across a wide range of generating, transforming, and distribution equipment (manufactured by ABB as well as others), such as changes in the temperature of transformers and secondary substations. This alerts utility control centers to possible overload conditions, allowing adjustments that can prevent blackouts before they occur" [12, page 2].

Data and messages can also flow from an instrumented smart object to a human such as data regarding vibration and harmonic readings and their trend over time, or a message from a machine about its health, or an indication on the defective parts that were produced in a machine on the shop floor of an IIoT manufacturing plant. Data and messages can move from a human to a machine for example giving an instruction or setting a threshold. Finally, data can flow among humans, for example, a Production Technician can send a request to the Maintenance Engineer that a piece of equipment is not working properly, a customer can provide feedback to the Quality Department on the satisfaction on the service being delivered, etc. All these data and messaging are delivered to each entity via a network.

As data, messages, and instructions are exchanged among the IIoT entities, the IIoT enterprise *Collects* and stores these data for future use. As mentioned earlier on, data can be structured (attributes and its values) or unstructured (text, video, and messages). Typically, the data collected in an IIoT system is time stamped, as it is generated through sensor monitoring or via transactions that occur in the IIoT ecosystem. To store the data collected, the IIoT organization uses a variety of data repositories and a *Data Lake* or a *Data Warehouse*. "The Big Data Lake/Data Warehouse has been conceptualized as a single data repository for an enterprise, typically designed to work in conjunction with Hadoop. Multiple sources such as data from an enterprise resource planning (ERP) system, sensor data from the Internet of Things, and data generated from Twitter or other sources would be in the same repository. Users would peruse and analyze the data as they search out solutions to existing and emerging problems" [11, page 70]. A Data Lake/Data Warehouse is a repository that contains raw data while a Data Warehouse is a data repository that has data that has been cleansed and pre-processed for analytic consumption in specific domain areas.

As the data collection activity proceeds, IIoT Entities and any application that interfaces with these entities use the data generated to *Learn* from the operational environment. The learning activity occurs by processing the data collected and identifying patterns in the data. The learning occurs when the system analyzes historical data and identifies patterns and trends and then utilizes these patterns to describe states, diagnose situations, project potential results, and predict future outcomes. Moreover, as new data is created as time moves forward, the algorithms can adapt and use these data to generate new and updated patterns (and hence learn from the environment) to then use these new patterns to aid domain experts in their decision making.

As IIoT entities use the data collected to learn patterns from the data, they then utilize the knowledge they have learned to *Do* or act to achieve or drive business outcomes. The "Do" activity in IIoT is the "actionable" element and takes many shapes. It can take the form of a message that a smart object sends to a user regarding its health (bearing #1 in the turbine is overheating). It can be an action that a smart object executes given certain circumstances (a robot fan increases it RPM's when a sensor in the controller reaches a temperature that indicates overheating). It can be a display in the operator's monitor of a frequent sequence of alarms and events when certain conditions are present, and the operator needs to take an action.

Instrumented smart products in the IIoT provide a great opportunity for new functionality, reliability, higher product utilization, scalability, and in one word transcend traditional product boundaries. IIoT organizations can create new businesses with the notion of smart interconnected elements.

A series of technological innovations are converging to make instrumented smart objects and products economically feasible. These technologies include high performance computing, electronics miniaturization, energy efficiency in batteries and instrumentation, advanced sensor technologies, low cost data storage repositories, economical computer processing power, cheap connectivity, wireless capabilities, Big Data Analytics, advanced visualization techniques, and powerful security protocols. Figure 1.3 shows a layered architecture depiction of an IIoT organization form both internal and external perspectives.

Layered View of IIoT systems

IIoT industrial organizations make instrumented smart products that are a combination of hardware, sensors, and software. Sophisticated production equipment, production methods, and business processes in IIoT organizations are required to create these instrumented and smart products. Hence, IIoT organizations need to (1) monitor and analyze internal operations that include production equipment, production processes, and business operations; and (2) monitor and analyze how their products behave when deployed at their customers operational environments.

An efficient IIoT organization then needs to use IIoT principles and operational concepts within the organization to ensure a smooth operation and use IIoT principles to develop smart interconnected products that offer its customers new ways to interact with the products they purchase. For this reason, the IIoT organization needs to build an infrastructure that supports both the internal and external elements of IIoT. Figure 1.3 shows an architectural view of the layers and components that need to be present in a fully operational IIoT organization, where the vertical layers with (1) show the internal perspective and the vertical layers with (2) show the external IIoT perspective.

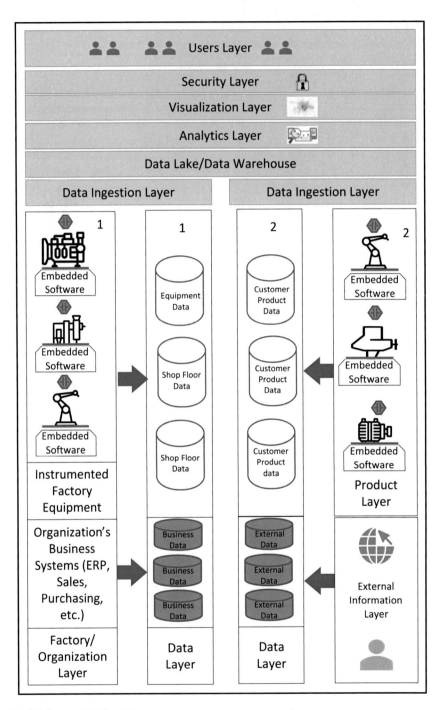

Fig. 1.3 Layered IIoT architecture

Starting from the left-hand side in Fig. 1.3, instrumented manufacturing equipment on the shop floor collects data from the manufacturing operations they perform, and data that monitors parameters associated with equipment health, maintenance data, and product and process quality data. Data that originates from the shop floor can be numeric, categorical, textual, pictorial and can be analyzed in real time at the equipment level (often referred to as "Edge Computing"[14]) at the shop floor level (often referred as "Fog Computing" [3]) , or in the Cloud (often referred to as "Cloud Computing" [1]).

These data are stored in local data repositories to either be analyzed at the shop floor level or to be extracted and ingested into a Data Lake or Data Warehouse of the IIoT organization and be analyzed at a later stage in the Cloud. The factory(ies) also generates business data (financial, inventory, sales, purchasing, customer complaints, etc.) that are stored in specialized system source repositories such as Enterprise Resource Planning (ERP), Purchasing, Customer Complaints, Customer Feedback (NPS), and others. These data can be numeric, categorical, and textual. The data that originates from IIoT factories and operations can have high volume of data, can be generated at high speed to be analyzed in real-time, and be of high variety (including not only numerical data but textual, pictorial, graphical, etc.), which describes the essence of Big Data concept [4]. An IIoT organization does not necessarily need to have all elements of Big Data at a point in time, but the IIoT platform needs to be able to handle and process Big Data. The data generated internally and stored in specialized data repositories is extracted and ingested by modules in the Ingestion Layer into the Data Lake or Data Warehouse as shown in Fig. 1.3. These data are continuously refreshed as it is generated and stored as raw data for future use. If an analytics application is defined, a specific data model is generated and from this data model the data attributes and associated data values are extracted into a partition in the Data Lake/Data Warehouse. Depending on the type of application, these data is refreshed at the required time periods. Once in this partition, the data is cleansed and pre-processed so that can be used for analysis and this partition is the curated data partition as shown in Fig. 1.3. To develop analytic applications, data scientists require a partition in the Data Lake that contains sample data extracted from the Curated Data partition to serve as a sandbox to experiment with algorithmic solutions and create the analytic processes required to implement a solution.

The Product vertical Layer (2) and the External Information vertical Layer are data generated outside the IIoT organization. The Product Layer data originates from instrumented products that the organization sells to its customers while they operate in their environment. These products are installed at customers sites and sensors capture data as the products do their productive work in the field capturing data related to specific parameters in the equipment (such as temperature, vibration and harmonics, gas concentrations, and other relevant parameters) as well as parameters associated with the production that these pieces of equipment perform (such as tolerances in parts processed, temperatures of mixes produced, viscosities, concentration of chemical elements, and others). These data are also stored in local data repositories and then ingested and transferred into the Data Lake for future analysis. Similarly, as in the case of data that originates on the shop floor of the IIoT organization, data generated at the customer site can also be analyzed in real or near-real time at the edge to for example, identify any anomalies in the products or in the process. External data can be included in the form of weather readings, or socio-economic indicators that can be combined with product data to derive sophisticated analytic correlations and predictions. The external and product data then become an important element of the new IIoT services that the organization offers to its customers. Another important source of external data includes data generated by customers. These external data originate from different customer layers. For example, data generated by customer financial people that explain reasons why a product or a service cannot be fully paid and reflects this information in the IIoT accounts receivable system. This type of data is typically transmitted from the customer's finance department as textual information. Another type of data includes customer feedback on their satisfaction on the products and services provided by the IIoT organization. Customer satisfaction data can be transmitted to the IIoT organization in the form of the Net Promoter Score (NPS) that includes a numerical customer satisfaction score and textual description of the customer satisfaction [13]. Yet another type of external customer data may originate from technical people at customers sites that provide feedback regarding anomalies as their equipment operates in the field. In the previous examples, it is evident that a lot of the external data generated by the different customer layers is textual and hence needs to be analyzed by natural language processing algorithms in the IIoT analytics layer.

The data that originates both internally or externally to the IIoT organization, can then be ingested by modules in the horizontal Data Ingestion Layer and stored in a Data Lake. Data in the Data Lake is initially stored as raw data. However, once a specific analytic application is identified, the data required is extracted and placed into a staging area where it is cleaned and pre-processed so it can be ready for analyses.

A brief note regarding data warehouses. In general terms, a data lake extracts raw data from original data sources without placing a lot of interest in filtering the original data. To use the data from a data lake, it needs to go through many pre-processing and cleansing operations. In contrast, data warehouses are typically used by enterprises to extract data in a less "indiscriminate" way and in a more targeted fashion. Hence, there is a lot of up-front analysis on the data tables and attributes that are extracted from the original sources. Although there is no best way to extract and store data, the data lake approach provides the opportunity to extract all data and then work needs to be done to cleanse it and pre-process it. A data warehouse can be viewed as a "targeted" river of data with a purpose for the organization and the structuring and data modeling work is conducted up-front. Nevertheless, is the opinion of this author that as data storage technologies continue to evolve, there is the possibility that both the data lake and the data warehouse concepts fuse to each other.

Once the data is stored in the Data Lake, the data can be analyzed from different perspectives, which are aligned with opportunities, pain points, problems, that the IIoT organization wishes to address. The analyses are performed utilizing a variety of analytic models made available in the Analytics Layer. Analytic approaches include statistical data analysis models, data mining models, machine learning models, and deep learning models. Chapter II in this book discusses a variety of analytic approaches utilized to analyze industrial data in the IIoT organization. A variety of analytic environments are available to data scientists to develop new algorithms or to re-use existing algorithms and they include Python, R studio, Watson, Azure ML MATLAB, etc.

The visualization layer is very important because it provides a user-friendly capability to show the users or domain experts the results of the analytics and to also provide the users with dashboards to "slice and dice" desired data partitions and analyses. Data visualization tools provide the designers with more convenient and faster way to create visual representations of large data sets. A variety of visualizations will be discussed and shown ads examples in upcoming chapters. The visualization layer can be provided using specific environments such as MS Power BI, Tableau, Grafana, and more. Tools such as D3 can be used to develop customized and specific visualizations that cannot be developed using tools such as Tableau or MS PowerBI. Data visualization techniques include charts (line, bar, or pie), plots (bubble or scatter plots), diagrams, maps (heat maps, geographic maps), and tables. All these techniques are available to user experience (UX) developers to provide the system users with a great way to view data and understand the results of analytics. An important element of the visualization layer is that in general terms tools such as Python or R possess a way to graphically show the results of complex analytic models. Nevertheless, by using visualization tools, the graphical representation provided by R or Python can be enhanced and made more understandable and accessible to the user domain experts.

In addition to all the above-mentioned layers, IIoT platforms must have a robust security layer that protects the whole environment. Developing end-to-end IIoT solutions involves multiple levels of security. Secure devices ensure security between users and devices (hardware). Secure communications ensure security in device-initiated connections and messaging control. Secure Cloud protocol involves security between the applications and the Cloud environment. Secure IIoT lifecycle management ensures security in remote control and updates of devices. Part of the security layer is the access protocols that allow different access levels depending on the user of the system.

The users of an IIoT system are knowledge workers, managers, Engineers, domain experts, and subject matter experts that interact with different components of the IIoT system. The role of these users is essential in an IIoT environment as they are the ultimate decision makers after the analytics are presented through the visualization layer.

Analytics Capabilities in IIoT Systems Can Increase Job Satisfaction

Analytics is the main engine of IIoT systems. Analytics is provided by machine learning algorithms, which are considered a branch of Artificial Intelligence. The machine learning and artificial intelligence capabilities in IIoT systems help to make commonplace industrial tasks more automated, more reliable, more consistent, and more accurate. Analytic methods also automate the analysis and computer learning capabilities of IIoT systems so that large volumes of historical heterogeneous data can be used to help human experts in their decision-making activities. The use of Machine Learning (ML), automation, and Artificial Intelligence (AI) techniques, which are essential in IIoT analytic systems, have caused many economists and authors to predict a hollowing-out of middle level, white collar jobs. Other authors instead, emphasize that the use of AI and ML in industry will create new jobs. Singh [15], concludes that low and middle skills level jobs will be reduced but highly skilled jobs where critical decisions are made using the results of AL and ML will increase. Huang and Rost [8] identify four levels of intelligence at which AI systems will progressively impact human labor, mechanical, analytical, intuitive, and empathic and eventually, jobs will be lost due to the efficiency of these AI systems in taking tasks in all four areas, although it will take time. When reading the literature hence, the opinion of authors seems to be split in terms of how the use on intelligent computing will either eliminate jobs or will create jobs. For the purposes of this book, the author presents below his perspective based on over a decade of developing advanced industrial analytic systems and considering the current state of the AI technology as applied to industrial applications.

Based on the current and future trends in industry it is possible to observe that important job skills include complex problem solving, critical thinking, synthesis of large volumes of data and information, identification of patterns and trends, and a high degree of creativity. We have observed that in the course of developing the IIoT applications described in this book, Artificial Intelligence (AI) approaches such as Machine Learning which is used for diagnostic, predictive, and prescriptive analytics in the IIoT, helps develop the future job skills (complex problem solving, critical thinking, and creativity) in domain experts needed to flourish in their jobs. Core activities carried out by subject matter experts (SMEs) such as Quality Managers, Inventory Planners, Account Receivable personnel, Service Engineers, Sales Portfolio Managers, Product Pricing Strategists, Health and Safety professionals, Service Engineers, and many others, require capabilities to analyze and visualize large amounts of heterogenous data very fast. This is not possible to be efficiently and comprehensively done by the human brain and this is where advanced analytics and machine intelligence play a crucial role to boost the analytic capabilities of these SMEs. AI techniques used in these cases, do not only help SMEs to do their jobs better, but also help the organization to hire new personnel to keep up with the work generated by analyzing this vast amount of data and information. Let us examine some specific examples below.

An original equipment manufacturing (OEM) organization often generates a stream of revenues by providing services to maintain their equipment in top shape. The service that OEM organizations provide include preventive maintenance, equipment diagnostics, proactive maintenance, and predictive maintenance. These services are mostly provided by highly specialized, expensive, and scarce Service Engineers/Technicians. The traditional service approach is extremely manual and focused on the capability and limited by the availability of these Service Engineers/ Technicians. Let us consider the case of a domain expert such as a Service Technician/Engineer that diagnoses possible causes why a machine may have failed on the field. In a traditional situation where the Service Technician/Engineer (or subject matter experts) does not have access to an IIoT environment, a large part of her/his work is dedicated to travel to the site where the equipment is installed (often remote sites), retrieve data stored in often obscure repositories, search for a segment of datapoints around the time where a malfunction occurred, try to visualize with limited tools the datapoints of interest, use her/his own expertise to diagnose a problem in the equipment, and propose potential solutions so that then the maintenance experts can fix the problem.

Compared with the traditional service approach, an IIoT organization with analytic service solution greatly enhances the service experience of both the Service Technicians and the customers. First, in an IIoT environment, a networked machine continuously sends sensor data to the IIoT system allowing the Service Technician to monitor in real or near-real time how the machine operates. This dramatically reduces the traveling requirements of the SME, eliminates the effort of tediously look for relevant data (as data is continuously monitored), and improves the visualization capability of relevant data in a user-friendly environment. This improves the work satisfaction of the SME, reduces costs (travel costs, time dedicated to extract relevant data, and utilize cumbersome tools to visualize relevant data). The cost reduction then allows the service organization to hire new Service Technicians that can service more customers with increased efficiency.

In the IIoT environment, the machine learning algorithms embody the domain knowledge on how to analyze the historical data to identify abnormal behavior of machines. It also allows for a standardization of the identification of problems, which will help more junior Service SMEs to perform complex diagnostics, and allow more senior SMEs to focus on more difficult problems, which will help increase the job satisfaction of these senior SMEs as they will use their skills and creativity to solve these more unique problems. Another advantage is that the knowledge in the algorithms is automatically updated with the new data generated. This allows the service offerings provided by the OEM to be more efficient, be more accurate, consistent, and less expensive to its customers.

Another example where we can see a positive impact of AI/Analytics in IIoT is in increasing the share of wallet of a company that has many businesses and sells products to a large variety of customers. In the traditional model, Sales Managers are assigned specific regions and have a portfolio of customers that typically buy certain types of products from the organization. The Sales Managers in general have a lot of experience and knowledge about the market, their customers and needs and habits, and the company's products. They then use this knowledge to maintain and even expand their sales, but in a way each Sales Manager often acts like a silo with limited communication with other Sales Managers and with little or no knowledge on customers outside their areas of influence due to a limited visibility on what is happening outside their limits. The power that advanced analytics provides an IIoT organization can be harnessed through an application that analyzes historical purchasing data (of both products and services) of its customers, identifies buying patterns, and compares similar customers across the whole organization to increase the share of wallet of the company. A system like this allows to optimize the work of Sales Managers and increase the volume of sales opportunities and hence, often, the need for more sales agents. This is another instance where analytic applications may result in generation of new jobs.

The section below describes few examples on how the Industrial Internet of Things is changing the business models of organizations that embrace this new approach. The examples described are real-world examples selected from different industries that summarize how the traditional business models were transformed using the IIoT concepts.

Examples of IIoT Business Models

The Industrial Internet of Things offers the opportunity to organizations to restructure their business models and change traditional revenue and economic models. This section describes several examples where organizations that embrace the IIoT philosophy can evolve their business models. Gilchrist [5] discusses couple of examples on how organizations can change their business paradigm by using the IIoT so, for example, the author points out that ultimately, consumers are interested in products in that they provide specific services or the so called "outcome economy". Customers are interested in a light bulb because it produces light so the light bulb manufacturer instead of selling the traditional light bulbs, in the IIoT framework would sell light/hours to its customers. Similarly, a logistics company would prefer to pay as they go for the use and consumption of tires, where the manufacturer instead of selling tires up-front would place sensors in the tires and charge for a mile of usage and consumption of the tires, which is an advantageous business model for both manufacturer and consumer. A summary of the IIoT business cases that will be reviewed in detail is presented in the subsections below.

Power Distribution Systems in the IIoT

Electricity required to supply electric power to cities, is produced in generating power stations, or power plants. Such generating stations are conversion facilities in which the heat energy of fuel (coal, oil, gas, or uranium), the sun, the wind, or the hydraulic energy of falling water is converted into electricity. The transmission system transports electricity in large quantities from generating stations to consumption areas. Electric power delivered by transmission circuits is typically considered as high voltage and must be "stepped down" in facilities called substations to lower voltages that are more suitable for use in industrial and residential areas. The part of the electric power system that takes power from a bulk-power substation to consumers, commonly about 35% of the total plant investment, is called "Power Distribution System". Power substations and their equipment play an essential role in the distribution of electricity. Much of the power distribution infrastructure in the US and other parts of the western world is over 50 years old. A key issue facing utilities is to efficiently utilize their limited funds for upgrade, maintenance, and repair of distribution lines. Studies in the UK show that more than 70% of unplanned customer minutes loss of electrical power are due to problems in the distribution grid [7]. For this reason, it is important to have prediction models that can foretell when a fault event may occur in a distribution network given certain conditions.

Work in the area of substation and power distribution system fault identification has mostly been reactive in nature, i.e. faults are identified and diagnosed after they have occurred. Fault diagnostics in substations and distribution systems has had limited automation capabilities. Since substation faults are likely to result in costly power outages, forecasting power system fault events before they occur can reduce response time, increase precision, and enhance preparedness to fix the outage; and all these reduce outage costs to both utilities and customers.

Power distribution grid operators currently rely on manual methods and reactive approaches to outage diagnostics and location identification. The efficiency of dispatching crews can be improved by having more automated diagnostic methods and fault predictive capabilities. According to a survey conducted by the Lawrence Berkeley National Laboratory, power outages or interruptions cost the United States of America $80 billion annually [6].

The advent of the IIoT technologies such as sensors, networked systems, robust and inexpensive data storage systems, wireless communications, powerful data analytics capabilities and visualization technologies provide a fertile ground to allow forecasting fault events and pin-point their location in a substation and power distribution grid to prevent expensive power blackouts. This can be done in both overhead power distribution lines and underground distribution lines. The IIoT allows companies that provide equipment to utilities to build their distribution grids to create new services aimed at utilities to make them more predictive in nature and less reactive and allow them to quickly isolate a power outage in a very narrow area in the grid. IIoT power distribution enabled systems also allow utilities to plan and be more prognostic when weather storms hit the power grid and hence reduce the duration of power outages. As a futuristic perspective, manufacturing organizations or OEMs that provide equipment to utilities, could change their whole business model and take the whole responsibility to provide utilities with "power hours" in their distribution grid and become fully responsible for the equipment and ensure power disruptions are dramatically reduced. Utilities can for example buy "Reliability" from manufacturers of power equipment similarly as airlines do not buy jet engines, but they buy reliability from Rolls Royce's TotalCare Gilchrist [5]. The impact on job satisfaction in this new business model in the utilities industry is important as with more precise knowledge on what type of failure and where is in the network, the "right" crews are dispatched to the specific location where the failure occurs. Moreover, with this predictive knowledge capability, power utilities can spend more time and budget hardening and upgrading the power distribution infrastructure spending more money hiring specialized crews to build a more reliable net reacting to failures.

IIoT in Process Control Alarm Management

Process industries use complex control systems to monitor process manufacturing operations. Control systems collect a large variety and volume of sensor data that measure processes and equipment functions during the production operations in industries such as refineries, oil and gas production facilities, chemical plants, power generation plants, etc. Alarms constitute an integral component of the data collected by control systems. These alarms are generated when there is a deviation from normal operating conditions in equipment and processes. With large number of alarms potentially occurring in a plant, it is imperative that operators and plant managers focus on the most important alarms and dismiss un-important alarms. Current alarm management systems provide basic statistical analysis and simple visualization capabilities to analyze historical alarm and event data stored in Control Systems' repositories. Current alarm management systems transform raw `alarm and event data to human-friendly and meaningful information supporting an operator's reporting and analysis needs, improving decision making processes, providing real time help at the alarm or event level. These systems allow users to conduct basic analyses of alarms such as determining the most frequent alarms in an event log, most common pair sequences of alarms, and basic analysis of nuisance alarms. Conventional alarm management systems help process industries to move from being overloaded and reactive with alarms, towards managing the alarm load and becoming stable in eliminating nuisance alarms and identifying most common alarms in the system to help the operators focus on most important alarms.

A new generation of intelligent alarm management systems are currently being developed leveraging the Industrial Internet of Things, Cloud computing and cyber-physical systems and this has introduced a promising opportunity to build powerful industrial systems and applications contributing to what has been called a "smart factory". With rapid advances in technology, sensors, actuators, networks, and data repositories are getting increasingly powerful and less expensive, which makes their use ubiquitous and very attractive for industries. As IIoT is getting widely applied to industries, machinery in industrial processes are getting instrumented with sensing, identification, processing, communication, analytics, and networking capabilities. The IIoT allows objects to be monitored and/or controlled remotely across existing network infrastructure. This opens opportunities for developing industrial applications for various needs such as automated and pro-active monitoring, control, decentralized decisions making, management, and maintenance resulting in improved efficiency, accuracy and economic benefits. These characteristics permeate the world of Industrial Control Systems and their Alarm Management sub-systems.

A new generation of alarm management systems is currently being developed in IIoT process industries where advanced analytic methods such as sequence mining, graph-based analytics, and market basket analysis are used to identify patterns of alarms and events that lead to deep reduction of nuisance alarms, predicting important sequences of alarms and events, and focus operators on making decision to address most important alarms. Additionally, novel visualization techniques allow operators in control rooms to view important alarm sequences in smart monitors. These new generation alarm management systems will allow providers to develop real-time monitoring systems for process plants where operators can visualize and interact with the system to reduce the potential for catastrophic accidents. Developers of control systems can then provide services to their customers that allow them to better manage their events and alarms and inform them when a critical chain of events may start occurring before a catastrophic event occurs. Intelligent alarm management systems greatly help plant operators in easily identifying areas that require immediate attention without being lost in analysis of nuisance alarms. IIoT organizations that implement intelligent alarm management systems can create and hold alarm rationalization workshops that can generate new job opportunities and help improving the operations of the IIoT plant and to improve the way control systems help the facility.

Power Generation Turbines Anomaly Detection

Electrical power generation plants are essential for the economy of our world. The main function of these plants is to generate electrical power 24 X 7 for consumption in cities, transportation infrastructure, and to power industries. Hence, it is essential that these plants operate at maximum efficiency and with minimal disruption. Power plants have equipment such as turbines, generators, stationary batteries, and electrical power is typically generated by rotating turbines that can be powered by gas, hydro, coal, wind, and nuclear energy. Power can also be generated by solar panels and inverter systems that transform solar energy into electrical power. All equipment in power plants needs to operate optimally to avoid costly electricity disruptions and turbines are not the exception. Turbines need to be continuously monitored and when breakdowns occur quick diagnoses need to be carried out to minimize operation disruptions. Turbines are very complex systems and as is very expensive to stop them from working for maintenance and therefore is important to have efficient monitoring methods and approaches. Manufacturers that provide equipment to power generation plants work in tandem with power plant maintenance operators to ensure turbines work properly and often monitoring of turbines is carried out within the power plant. Turbines are typically retrofitted with a large number and variety of sensors placed in strategic sub-components of the turbine such as bearings, coils, axels, generators, etc. These sensors measure a large variety of parameters such as vibrations, temperatures of components, ambient temperatures, fluid temperatures, electricity outputs, etc. These data are continuously collected typically at very short periods of time (minutes) and stored in the power plant's local control system repository. The monitoring capability in these situations occurs at scheduled intervals of time or if the operators recognize an obvious misbehavior in the turbine (such as abnormal vibrations or noise, drop in the power generated, etc.). The typical procedure then is that if the operator at the plant has observed a mis-behavior that cannot diagnose, he/she calls a service person from the OEM or equipment provider and the service person physically downloads the data from the control system into a laptop computer and then analyzes the data using his/her own methods and domain expertise to diagnose the mis-behavior. Still, nowadays there are many power plants that operate this way despite the advent of the IIoT. This way of operation makes the maintenance of turbines and similar equipment in power plants extremely reactive.

An IIoT enabled power plant and equipment provider operates in a more efficient manner to address anomalies in the operation of a turbine or other equipment. The data gathered by sensors is transmitted into the power plant control system and then ingested into the IIoT Data Lake utilizing the IIoT secure network. The equipment provider then has the capability to continuously monitoring the behavior of the turbine (or equipment) by utilizing powerful data mining and machine learning algorithms stored in the analytics layer of the IIoT platform. Nowadays, this approach allows the equipment provider to also furnish services to power plants remotely and in real-time. This capability dramatically increases the probability to detect early on anomalies that if left alone could bring down the turbine and cause power disruption. An IIoT anomaly detection system then allows the customer to reduce the probability of equipment failure and hence sever economic losses, and allows the equipment provider to furnish preventive, proactive, and even predictive services to its customers.

As explained earlier in the chapter, full embracement of IIoT brings a potential to fundamentally change the business model between equipment providers and customers. In this case, for example, the turbine (equipment) provider could work with its customers and instead of selling the equipment sell hours of power generated, or even sell the power plants number of megawatts per year and become fully responsible to monitor, maintain, diagnose, and service the equipment and the customer be focused only in its downstream customers and not be concerned about the equipment.

From the job creation perspective, a remote and intelligent anomaly detection system helps reducing the travel costs and those savings facilitate hiring new service personnel to help more customers. An intelligence remote monitoring system allows the OEM and service provider to hire many Service technicians that may not need to have decades of experience and use the highly specialized domain experts to work on very difficult problems.

Increase Share of Wallet of Industrial Services and Products

Global industrial organizations that operate in many countries around the world, have many diverse businesses, offer a large variety of products and services, and their businesses are geographically dispersed face a complex challenge in identifying opportunities to up-sell, cross-sell, and new sells of products and services. Often, due to its distributed nature, it is almost impossible to understand the purchasing patterns of their globally distributed customers. Similarly, the businesses themselves may be operating and selling products and services using different approaches in different geographic regions. So, it is important for these large conglomerates to find a coordinated way to increase the share of wallet of their customers and to identify opportunities to increase sells of products and services where gaps in portfolio exist. Additionally, a system that provides real-time visibility on these selling opportunities is of great value.

The main objective then is to have an IIoT system capable of analyzing sales and service data originating from the global sales of product and services and execution of services stored across various systems such as Customer Relationship Management, Enterprise Resource Planning, and Service Support systems to efficiently monitor various KPIs related with product sales, service sales, service execution, customer purchase, customer purchase preferences, installed base management, customer demographics, customer satisfaction. The analytic approaches that are implemented are expected to derive key performance indicators (KPI's) that are predictive in nature and are derived using data mining and machine learning algorithms. Ultimately, the objective is to have a Machine Learning-based system that provide recommendations to Sales Domain Experts on new products and services that can be sold to their customer base and are not currently buying them. The system functionality aims at analyzing the existing set of customers who are regularly engaging with the organization and buying specific products and services (the company's customer base), to identify what might be other products or services to be sold to these customers.

An IIoT system capable of identifying potential new sales of products and services to customers in a globally distributed environment has the potential to change the business model of an IIoT organization in several ways. First, it creates a common consistent way for the multinational organization to address all its customers that are geographically dispersed and increase customer satisfaction not only locally but globally. Second, it allows the different businesses in the multinational organization to learn about the successes of interactions in a geographic region and expand it in all geographic regions increasing both the profitability of the organization and the customer satisfaction. Third, the IIoT system allows the product and services salespeople across different businesses and geographic regions in the organization to learn about each other, learn about their products, and learn about their customers. Fourth, it opens a potential to sell products and services not totally new customers that the system may deem like existing customers. Finally, it allows the IIoT multinational organization to be proactive, predictive, and prescriptive rather than reactive to its markets and customers globally. From the job perspective, salespeople can be more productive and increase the revenues of the organization. With these new revenues, new sales channels open and hence new salespeople are needed to address these new markets.

Power Transformers and Utility Equipment Analysis

Electric utilities face a continuously growing number of challenges to reliably maintain and operate a complex power and distribution grid, maintain a hybrid power grid composed of outdated and highly sophisticated systems, deal with weather-related challenges and adapt to customers' power requirements that need to satisfy change at a speed never experienced before. Major constraints imposed by an aging and not adequately replaced workforce, aging assets, newly created operational and compliance requirements, distributed generation, new types of load (e.g. electrical vehicles), very fast communication channels co-existing with some still very old electromechanical devices (e.g. relays), the unspeakable amount of new data which is daily generated and stored, and high interconnectivity are certainly some of the huge challenges to be faced by the existing operational and maintenance procedures.

Utilities are responding to these challenges in a variety of ways. On the hardware side there are several new power system technologies available, from more compact and reliable devices to the application of sophisticated online sensors, providing a new suite of data which should ideally help in the determination of, for example, asset operating condition. The introduction of new communications infrastructures into existing and outdated substations and transmission lines, as well as the construction of fiber optics data links and new data centers are shaping a new paradigm in the electrical industry, addressing a number of issues but still lacking response to a few others. Recent trends in the central acquisition and storage of company-wide data, thus facilitating not only the archival but also the analysis of historical information and hopefully contributing to a more reliable and less costly asset management and systems operations.

IIoT platforms and software tools are capable of handling large amounts of old and new data which are generated by power system assets such as power transformers. IIoT helps to incorporate expert knowledge into the existing or new asset management tools, thus addressing the following topics of paramount importance: a) simultaneous analysis of online data (from sensors) and offline data (from manual tests - often performed while assets are removed from operation) b) multi fold data acquisition rates, from annual or monthly to continuous streams of data c) automated systems incorporating expert knowledge regarding individual assets d) adequate reporting through multi-persona dashboards, with key performance indicators as well as detailed technical information and recommendations e) feedback incorporation of human actions and maintenance information into the analysis.

Although the example given below refers exclusively to power transformers, the same principle may be applied to the construction of similar structures of other assets. The above-mentioned elements are part of the IIoT infrastructure being implemented in the utilities industries. This infrastructure is conducive to continuously monitor equipment in the utilities, especially costly equipment such as power transformers.

Manufacturers of power equipment or OEMs can then monitor in real-time their utilities customers' equipment operating in the field and identify mal functions that can lead to costly breakdown of equipment. Through the IIoT power equipment manufacturers are helping utilities to prevent these costly breakdowns and at the same time generating a new revenue stream of services that are less reactive and more prognostic or predictive. In the future, power equipment providers could change their business model and instead of selling the physical equipment they could sell number of hours of continuous operation of this equipment and become fully responsible for the equipment, so it works without stopping. Not only that, but they can reduce the number of physical configurations of this equipment and via software, offer different options (such as power values, etc.) to utilities that can activate via the cloud and enable the equipment to meet their requirements. By opening a new service capability, the OEMs need to hire new personnel and Service Technicians to keep their customers satisfied.

Demand Forecast of Products and Spare Parts

An important activity conducted in production planning in factories is demand forecasting and it is used for production planning and determining inventory levels in the factory. The IIoT factory is highly interconnected internally and externally with customers and suppliers and typically sells a large variety of products all over the world and it needs to track all transactions efficiently. The IIoT factory needs to respond to changes in customers' demands and suppliers' constraints very fast and efficiently. It is important then for production planners and inventory management personnel in the IIoT factory to have access to adaptable, fast, and sophisticated demand forecasting system that allows them to create a variety of demand forecasting scenarios to ensure all parts needed for manufacturing are available and to ensure demand of the customers is fulfilled. IIoT factories need to have a demand forecasting system that provides them with a large variety of forecasting algorithms that cover all spectrum of demand profiles. This will allow for the development of a variety of demand scenarios that can be used to adapt to changes in the market and in suppliers. To run this demand forecast scenarios, typically IIoT organizations use Big Data approaches that allow running these algorithms efficiently and with high performance.

References

1. Botta, A., de Donato, W., Persico, V., & Pescape, A. (2016). Fog computing: Helping the internet of things realize its potential. *Elsevier Future Generation Computer Systems, 56*, 684–700.
2. Chou, T. (2016). *Precision: Principles, practices and solutions for internet of things.* Cloudbook, Inc., USA
3. Dastjerdi, A. V., & Buyya, R. (2016). Fog computing: Helping the internet of things realize its potential. *IEEE Computer, 49*(8), 112–116.
4. De Mauro, A., Greco, M., & Grimaldi, M. (Eds.). (2016). A formal definition of big data based on its essential features. *Library Review, 65*(3), 122–135.
5. Gilchrist, A. (2016). Industry 4.0: The industrial internet of things. Apress Editors. Apress, Berkeley, CA
6. Hamachi, L. K., & Eto, J. (2004). *Understanding the cost of power interruptions to U.S. electricity consumers.* OSTI.GOV, US Department of Energy, Technical Report LBNL-55718, September 1st.
7. Hampson, J. (2001). Urban network development. *Power Engineering Journal, 15*(5), 224–232.
8. Huang, M.-H., & Rust, R. T. (2018). Artificial intelligence in service. *Journal of Service Research, 21*(2), pp. 155–172.
9. Madakam, S., Ramaswamy, R., & Tripathy, S. (2015). Internet of Things (IoT): A literature review. *Journal of Computer and Communications, 3*, 164–173.
10. Manu, A. (2015). *Value creation and the Internet of Things: How the behavior economy will shape the 4th industrial revolution.* London/New York: Routledge Taylor and Francis Group.
11. O'Leary, D. (2014). Embedding AI and crowdsourcing in the big data lake. *Intelligent Systems, 29*(5), 70–73.
12. Porter, M. E., & Heppelmann, J. E. (2014). How smart, connected products are transforming competition. *Harvard Business Review*, November issue.
13. Reichheld, F. F. (2003). One number you need to grow. *Harvard Business Review*, December issue.
14. Satyanarayanan, M. (2017). The emergence of edge computing. *IEEE Computer, 5*(1), 30–39.
15. Singh, D. and Geetali (2019). *Employment transformation through artificial intelligence in india.* International Journal of Applied Engineering Research, ISSN 0973-4562, 4(7) (Special Issue).

Chapter 2
Industrial Analytics

Industrial analytics refers to the analysis of large variety and large volume of historical data generated by industrial organizations during their operations, with the objective of improving internal and external decision making. The Industrial Internet of Things provides a vehicle to utilize historical data generated in the organization and utilize analytics methods to find important answers that bring economic benefits to the organization [1].

Although the literature presents many options to classify data analytics algorithms, for the purpose of this book data analytics is classified into four main categories: (a) Descriptive Analytics; (b) Diagnostic Analytics; (c) Predictive Analytics; (d) Prescriptive Analytics.

Descriptive analytics refers to approaches that analyze historical data using Statistical methods that summarize historical data and convert it into a form that can be easily understood by domain experts. Descriptive Analytics explain in detail events that have occurred in the past. This type of analytics identifies data patterns and trends observed from past events and draws interpretations from them so that sound strategies can be developed for decision making. Often, this represents the initial type of analytics that is conducted in an IIoT organization and results in basic key performance indicators in an organization. Descriptive analytics answers the question of what happened in the past. Descriptive analytics findings signal that something is wrong or right, without explaining why. For this reason, highly data-driven companies do not content themselves with descriptive analytics only and prefer combining it with other types of data analytics. Moreover, visualizing data in dashboards that allows to filter, slice-and-dice these data is a common descriptive analytics approach that organizations implement at the beginning of their IIoT journey. This is a very important step for organizations as it provides the capability to store and view from one location (the IIoT data lake or data warehouse) data that otherwise may not available.

Diagnostic analytics aid domain experts to explore deeper into an issue at hand so that they can arrive at the source of a problem and answer the question of why something happened. With diagnostic analytics there is a possibility to drill down,

A. Dagnino, *Data Analytics in the Era of the Industrial Internet of Things*,
https://doi.org/10.1007/978-3-030-63139-0_2

find out dependencies and identify interesting patterns using historical data. Diagnostic analytics provides in-depth insights into a problem. In this approach, historical data is used to help determining the cause of a malfunction, an anomaly, or an irregularity in a physical or socio-economic system being studied or analyzed. Diagnostic analytics is closely integrated with domain knowledge of subject matter experts (SMEs), as they typically guide how the analytics process is carried out to identify these anomalies that are built based on domain expertise. Often, the SMEs and data scientists need to work together to develop the analytics process and methods that follow the domain expertise roadmap.

Predictive analytics helps businesses to forecast, project, or predict trends based on past events. Usually, many different but co-dependent variables are analyzed to predict, project, or forecast a trend in this type of analysis. For example, in the healthcare domain, prospective health risks can be predicted based on an individual's habits/diet/genetic composition. Predictive analytic models are very important across various fields in different domain such as industry, manufacturing, health care, etc. Predictive analytics tells what is likely to happen. Predictive analytics use various machine learning algorithms that can individually consume historical data or can combine the findings of descriptive and diagnostic analytics to detect tendencies, clusters and exceptions, and to predict future trends, which makes it a valuable tool for forecasting. Despite numerous advantages that predictive analytics brings, it is important to understand that forecasting or projection represent just an estimate, the accuracy of which highly depends on data quality and stability of the situation, so it requires careful consideration by the domain experts and continuous improvement.

Prescriptive analytics prescribes domain experts what action can be taken to eliminate a future problem or take full advantage of a promising trend. Prescriptive analytics also uses advanced tools and technologies, like machine learning, business rules and algorithms, which makes it sophisticated to implement and manage. Figure 2.1 shows a diagram that represents the four types of analytics in a Venn-type diagram.

Industrial Analytics is defined in this book as a way of analyzing and inspecting a corpus of selected historical data of an IIoT organization and it is primarily used to make data-driven decisions that are actionable. Industrial Analytics is all about gathering data inputs all around the organization and rendering analytics to improvise future performance. Industrial Analytics often drives towards data-driven predictions.

Another important concept in industrial analytics is data mining, which refers to extracting knowledge from a large amount of data and is the process of discovering various types of patterns and trends that are inherited in the historical data and are useful for decision making.

Machine learning is also a very important concept in industrial analytics and it refers to the utilization of algorithms that use historical data to automatically generate knowledge and improve it through experience based on the historical data [1]. Machine learning involves the study of algorithms that can extract information automatically from a corpus of historical data [3]. Machine-learning uses data mining

Fig. 2.1 Types of analytic approaches

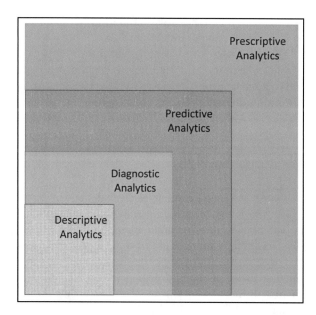

techniques and another computerized learning algorithms to build models of what is happening by using historical data so that it can predict, forecast, project, or prescribe future outcomes.

From the above discussion, we can see that Statistical Analysis, Data Mining, and Machine Learning approaches work together to provide the IIoT organization with a series of data analytic processes (made available through the Analytics Layer) that drive data-driven decision making. Statistical Analytic models analyze historical data and are fundamental to Descriptive and Diagnostic Analytics. Data Mining algorithms analyze historical data and identify patterns and trends in the phenomena being analyzed and are fundamental for Diagnostic Analytics and bridge Predictive Analytics. Machine Learning algorithms use the patterns discovered in the Data Mining Process and use these patterns as a basis to predict or prescribe solution to problems and use new data to learn and adapt to changes in the environment. Machine Learning may use Data Mining techniques to build models and find patterns, so that it can make better predictions. Data Mining can use machine learning techniques to produce more accurate analyses and identify trends in historical data.

Data mining can be a more manual process that relies more on human intervention and decision making. In Machine Learning, once the initial set of rules are in place, the process of extracting information and 'learning' and refining using new data is performed automatically and this takes place without or with minimal human intervention. In Machine Learning the algorithms become more intelligent by learning from new data.

Data mining is used on an existing dataset (like a data lake or data warehouse) to find patterns. Machine Learning algorithms, on the other hand, are trained on a 'training' data set, which teaches the computer how to make sense of data, and then they make predictions on a subset of historical data and used to make prediction on totally new data.

For the future discussions in this book, the focus will be on describing how Machine Learning and Statistical models can be used as descriptive, predictive, forecasting, and prognostic analytic approaches to solve important problems in Industrial Internet of Things organization.

Machine Learning

Although Machine Learning has been defined in many ways, for the purpose of this book, we define Machine Learning as a computing capability that assists in the autonomous acquisition and integration of knowledge using a corpus of historical data. Machine learning is useful when data is continuously retrieved in rapid changing phenomena, for example in industry reading sensor data from equipment, production, diagnosis, quality, demand, sales, etc. Machine learning is also useful when there is the need to customize solutions for specific smart objects or the knowledge worker. Machine learning brings together several disciplines such as data mining, statistics, control theory, cognitive science, artificial intelligence, psychological models, evolutionary models, database theory, and information theory.

Machine learning algorithms build a mathematical model based on sample historical data, that is known as "training data" to make decisions without being programmed to perform the action [1]. Machine Learning algorithms can be classified into (a) Supervised Learning; (b) Unsupervised Learning. Figure 2.2 shows the two primary branches of machine learning with their associated algorithms. The green boxes show the algorithms that are explained and used in the examples described in this book. The grayed boxes are algorithms not covered in examples in this book.

Supervised Machine Learning

Supervised Learning refers to algorithms that teach the computer patterns and trends using labeled historical data [11, 12]. In Supervised Learning algorithms, instances are provided with known labels (that are the corresponding correct outputs). Two main categories of Supervised Learning algorithms can be identified: (a) Classification; (b) Regression.

Classification is a supervised learning approach where the computer learns from a data input given to it and then uses this learning to classify a new observation. Supervised Classification algorithms identify the category that a new observation belongs based on a training set of data containing observations (or instances) whose category membership is known [11]. Figure 2.2 shows that within the Classification Machine Learning the following algorithms are included: Decision Trees, Random

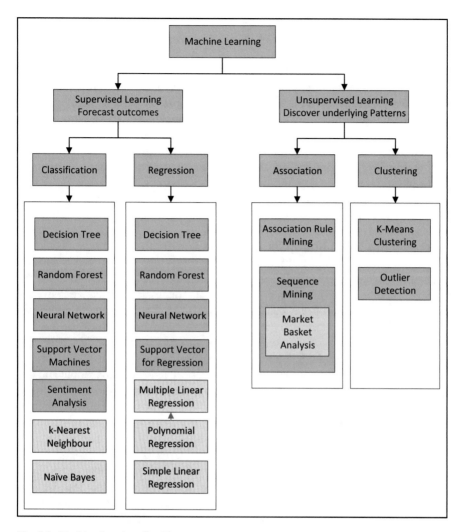

Fig. 2.2 Machine learning algorithms

Forest, Neural Networks, Sentiment Analysis Algorithms, Support Vector Machines, k-Nearest Neighbour, and Naïve Bayes.

Regression is also a supervised learning approach that refers to systems that predict a value for an input based on previous information. In Regression an input is mapped to an output based on earlier observations. But in the case of regression the algorithm predicts a value and not just a class of an observation. Two important characteristics of regression are that the responses that can be expected from the model are always quantitative in nature, and the model can only be created by taking into consideration past data. Figure 2.2 shows that the following algorithms are included within the Regression Machine Learning algorithms: Decision Trees for

Regression, Random Forest for Regression, Neural Networks for Regression, Simple Linear Regression, Multiple Linear Regression, Polynomial Regression, and Support Vector for Regression.

Although there can be many ways to align ML algorithms, this chapter provides a categorization based on how these algorithms can be used in industrial applications such as the IIoT. Figure 2.2 shows those algorithms that are used in the examples presented in this book shaded in green. For these algorithms, a detailed description is given that outlines in detail how these algorithms operate on historical data.

Some of the descriptions provided were inspired on Emerenko, K. explanations in his Udemy course entitled " Machine Learning A-Z: Hands-on Python and R in Data Science.

Decision Trees for Classification and Regression

Classification and Regression Trees (CART) refer to Decision Tree algorithms that can be used for classification purposes or for regression predictive purposes. Classification Decision Trees can use both categorical and numerical data while Regression Trees use numerical data.

Classification Decision Trees are a type of supervised Machine Learning algorithms that have a pre-defined target variable and it is used to solve classification problems. This algorithm works for both categorical and continuous input and output variables. Classification Decision Trees split the population or sample into two or more homogeneous sets (or sub-populations) based on most significant splitter / differentiator of input variables.

The Classification Decision Tree is a tree like collection of nodes intended to create a decision on values affiliation to a class or an estimate of a numerical target value. Each node represents a splitting rule for one specific variable or attribute. For classification, this rule separates values belonging to different classes. The building of new nodes is repeated until the stopping criteria are met. After generation, the decision tree model can be applied to new examples and each example follows the branches of the tree in accordance to the splitting rule until a leaf is reached.

The Given two independent variables X_0 and X_1 represented in a two-dimensional graph and a dependent variable (represented as a third dimension as 0 - green color or 1 – red color), Fig. 2.3 shows a set of points in the two-dimensional space. Notice that points have three values in the graph (X_0, X_1, and green or red color). The color denotes binary value of either green or red that denote the value of the third dimensional variable.

The Classification Decision Tree algorithm progressively sub-divides the space set into partitions or areas that contain majority of similar points. In this simple example, the first partition is set at a value of $X_1 < 30$ (or below the value of 30). The algorithm identifies major block concentrations of points above $X_1 < 30$ and below this threshold. This first partition becomes then the first node in the tree structure shown in Fig. 2.4. After partition 1, the algorithm decides to focus on the values above $X_1 < 30$ and it sub-divides the space using the value $X_0 < 40$ into two homogeneous segments. As can be observed, the segment on the right of $X_0 < 40$ is pure with red points, while the segment on the left of $X_0 < 40$ is almost pure with only

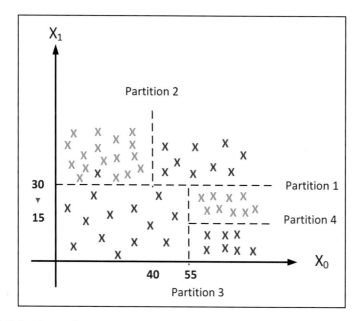

Fig. 2.3 Points in two-dimensional space and decision tree partitions

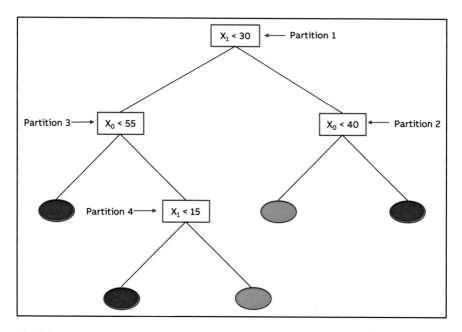

Fig. 2.4 Classification decision tree

one observation green. This is a natural impurity that the algorithm cannot totally avoid. Partition 2 then becomes a second node in the tree structure shown in Fig. 2.4.

As the algorithm continues its computations, the next partition (Partition 3) identified is at $X_0 < 55$ that divides the space area into two areas. The area at the left of $X_0 < 55$ contains a homogeneous sub-set of red data points while the area on the right contains a combination of green and red data points. Partition 3 is represented also as a node in the tree in Fig. 2.4. Finally, the algorithm identifies a new partition at $X_1 < 15$ that sub-divides the space into two homogeneous areas, the area above the partition with green data points and the area below with red data points. Partition 4 is also represented in the tree structure in Fig. 2.4.

The above explained dataset and partitions can represent any industrial, social or economic phenomenon. For example, X_0 could be a measurement of vibrations while X_1 can be a measurement of temperatures in a bearing. The third variable Y can take a binary value such as "normal" (green) and "anomaly" (red). In our example, if a new data point with coordinates ($X_0 = 27$, $X_1 = 70$) is read into the system coming from sensors, the trained Decision Tree Classifier model determines the value of the Y coordinate based on the area where the new point lands in the coordinate space to be in the top left hand-side area as green or normal.

When using Decision Trees data scientists must be aware that these models have a tendency of overfitting. The concept of overfitting refers to the creation of an analysis that fits too closely to a particular dataset and may therefore fails to predict or classify future observations reliably [9].

The **difference between** the **Classification Decision Tree** and the **Regression Decision Tree** is their dependent variable. **Classification Decision Trees** have dependent variables that are categorical and unordered. **Regression Decision Trees** have dependent variables that are continuous values or ordered whole values.

Random Forest Classification and Regression

Random Forest or Random Decision Forest is an ensemble learning algorithm that uses Decision Trees for classification and operates by constructing a multitude of decision trees at training time and generates a class that is the mode of the classes (classification) of the individual Decision Trees [10, 11]. Random decision forests address the susceptibility that Decision Trees have on overfitting the training data. Ensemble ML models, such as the Random Forest Algorithm, refer to combining the results of other ML algorithms into one model to optimize the results [14]. Random Forest leverages the power of the crowd to obtain the best result.

There are four basic steps that a Random Forest Algorithm executes to be created. These steps are explained below.

1. Select a random number of data points from the dataset.
2. Build a Decision Tree associated with the randomly selected data points.
3. Select the number of trees to be built and repeat steps 1 and 2.

4. For a new data point, use all trees to classify the data point and assign the new data point to the category that wins the majority vote.

A Random Forest model works well as a single decision tree may be prone to a noisy solution but aggregate of many decision trees reduce the effect of noise giving more accurate results.

The **difference between** the **classification random forest** algorithm and the **regression random forest** algorithm is their dependent variable. **Classification random forest** has dependent variables that are categorical and unordered. **Regression random forest** algorithm has dependent variables that are continuous values or ordered whole values. In case of **classification random forest** algorithm we try to classify new data entry based on maximum number of votes or mode of the output we get from each of the decision tree, whereas in Regression we assign the new data point the average across all of the predicted values from decision trees.

Neural Networks for Classification and Regression

Deep learning is possible due to high storage capacity and powerful processing power of computers. A promising method to store even more data than current silicon technology is DNA digital data storage that is the process of encoding and decoding binary data to and from synthesized strands of DNA (Wikipedia references). While DNA as a storage medium has enormous potential because of its high storage density, its practical use is currently severely limited because of its high cost and very slow read and write times. (Wikipedia references). Processing power is governed by Moore's Law. Geoffrey Hinton is the Father of Deep Learning.

Figure 2.5 shows a graphical representation of an Artificial Neuron that is shown as the green node. The green neuron or node receives input signals or values and output signals or output signal or value. Input signals are represented with orange nodes (or neurons) and output value is represented as a red node (or neuron). The yellow neurons are the independent variables and are called the Input Layer neurons, while the green neuron is in the so-called Hidden Layer. Notice that the input values are associated with one observation in the system or raw in the matrix. The independent variables need to be standardized or normalized to get them with values between zero and one so that the ANN processes the values easier [13]. The output value could be continuous numerical, binary, or categorical. Same as with the input variables, the output variable corresponds to the single output associated with the input variables in the Input Layer.

Weights are crucial to the ANN and they are adjusted through the ANN process of learning and moving towards the output. Inside the Node, a function is assigned to the weighted sum of values multiplied by the assigned weights and the result is passed on as an output value to the red node. In summary, the first step is to pass input values to the Green Neuron, in the second step the input values are processed by the Green Neuron utilizing a function of a weighted sum, and in the third step the resulting value is passed on to the red neuron.

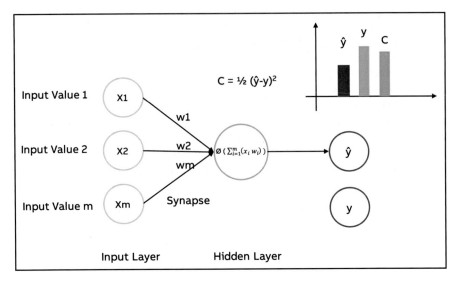

Fig. 2.5 Artificial neural network

Neural Networks use activation functions are used to determine the firing of neurons. Given a linear combination of inputs and weights from the previous layer, the activation function controls how information is passed on to the next layer. An ideal activation function should be nonlinear and differentiable. The nonlinear behavior of an activation function allows the neural network to learn nonlinear relationships in the data. The differentiable component allows to backpropagate the model's error when training to optimize the weights in the model.

There are several activation functions that are used in Neural Networks and these include: (a) Hyperbolic Tangent; (b) Inverse Tangent; (c) Rectified Linear Unit Function (very used in ANN); and the Leaky Rectifier Linear Unit. For more details the reader is referred to Glorot et al. [7]. These four types of activation functions can be visualized below in Fig. 2.6.

The function for the Hyperbolic Tangent Activation Function is:

$$f(x) = \frac{e^x - e^{-x}}{e^x + e^{-x}}$$

The function for the Inverse Tangent Activation Function is:

$$f(x) = \tan^{-1}(x)$$

The function for the Rectified Linear Unit Function is:

$$f(x) = 0 \; for \; x < 0$$

$$f(x) = x \ for \ x \geq 0$$

The function for the Leaky Rectified Linear Unit Function is:

$$f(x) = 0.1x \ for \ x < 0$$

$$f(x) = x \ for \ x \geq 0$$

To train the neuron shown in Fig. 2.5 the model uses the inputs x_i and multiplies them by the weights w_i to derive $\hat{y}$. The value of $\hat{y}$ is then compared to the actual y value and in the first iteration there will probably be a discrepancy.

This discrepancy can be measured with a cost function shown in Fig. 2.5 as C. The objective then is to feed the value of C back into the network and adjust the values of w_i and re-calculate a new value of $\hat{y}$ and iteratively converge to the value of y as the new updated values of w_i provide this convergence. Once this state is reached, then the Artificial Neural network is trainied and can be used to analyze new data.

Figure 2.6 shows an example on how a NN functions. In a trained NN, each node of the Hidden Layer calculates using a selected activation function a different combination of inputs and produces the $\hat{y}$ result as shown. Each node in the hidden layer sends its output to the Output Layer node to calculate the final value of $\hat{y}$. Classification artificial neural networks seek to classify an observation as belonging

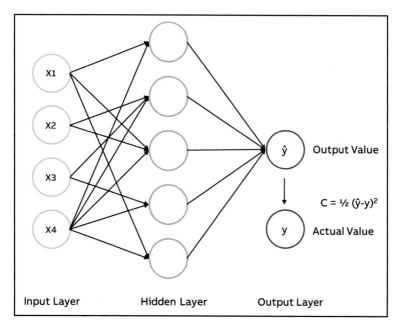

Fig. 2.6 Example on how a NN works

to some discrete class as a function of the inputs. The input features (or independent variables) can be categorical or numerical types, however, we require a categorical feature as the dependent variable.

Sentiment Analysis and Machine Learning

Sentiment Analysis is the ability of using Natural Language Processing and Text Mining machine learning methods to identify and extract attitude and emotion from a corpus of text. Sentiment analysis determines the degree of which a textual expression is positive, neutral, or negative. This is a field of Machine Learning that attempts to analyze and measure the human emotion behind a piece of text so that it can be transformed into useful business intelligence [15]. Sentiment analysis analyzes the value of the words in a piece of text and provides insight into the emotion behind the words. Sentiment analysis provides scores to a piece of text to determine the emotion behinds its words. Figure 2.7 shows the Sentiment Analysis process.

Analyzing data refers to the extraction of text data and textual and non-textual data are separated. Non-textual data is eliminated. Sentences are examined for "subjectivity" content and "subjective" sentences are retained. Subjectivity refers to the linguistic expression of someone's opinions, sentiments, emotions, evaluations, beliefs, and speculations on a subject [4]. For example, a customer may express: "I am very happy with the responsiveness of my supplier to rectify the missing parts in their last shipment".

During the indexing phase, sentiment sentences are classified into positive, negative, neutral, good, bad, like, dislike. Sentiments can be positive, negative or neutral. Unstructured text data is then converted into meaningful information. The ML approach to sentiment analysis is usually modeled as a classification problem where a classifier algorithm is fed with text and it returns the corresponding category as "positive", "negative', or "neutral".

The sentiment analysis algorithm and process are depicted in Fig. 2.8. In the training phase, the sentiment analysis model learns to associate a particular input (from the corpus text) to the corresponding output (label) based on the test samples used for training. The feature extractor transfers the text input into a feature vector. Pairs of feature vectors and labels (e.g. positive, negative, or neutral) are fed into the machine learning algorithm to generate a model. In the prediction phase, the feature extractor is used to transform unseen new text inputs into feature vectors. These

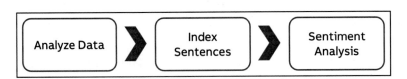

Fig. 2.7 Sentiment analysis process

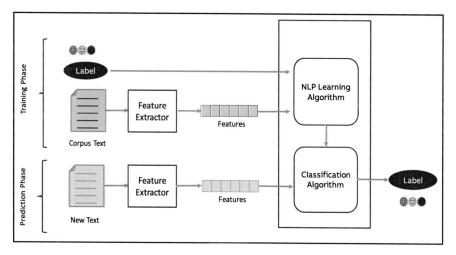

Fig. 2.8 Sentiment analysis training and prediction

Fig. 2.9 Clusters of points that are to be classified by an SVM algorithm

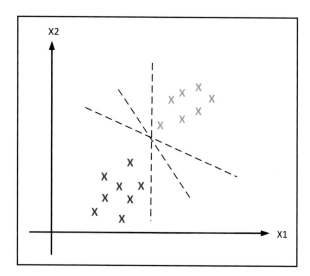

feature vectors are then fed into the model (through the classifier), and then generates predicted labels on the new text (again, positive, negative, or neutral).

Support Vector Machines

Let us consider the points in a two-dimensional space X1 and X2 as shown in Fig. 2.9. The objective is to a line that separates both point clusters so that when a new point is considered, it can be correctly classified. There can be a lot of lines that

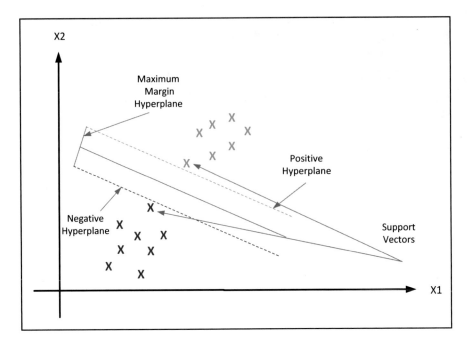

Fig. 2.10 Line separation created by SVM algorithm

can be drawn to separate these two clusters and depending on the choice, a new point may lay in either one point or the other. The objective of the Support Vectors Machine is to identify the best line(s) that separate both clusters.

The support vector machine algorithm identifies a line that is equidistant to both red and blue points and the sum of the distances between the line and each of the blue and red points (Maximum Margin Hyperplane in a multi-dimensional space) are maximized. Both red and blue extreme points are called the support vectors in the multi-dimensional space, and the remaining points do not contribute to the results of the algorithm. These two points are supporting the decision boundary. The line in green is called the positive hyperplane and the line below in red is called the negative hyperplane. What makes the SVM algorithm special is that it selects the most extreme points to set as support vectors which are very close to the boundaries. It is a more extreme algorithm as it looks at the most extreme cases and creates the boundaries at these extreme points (Fig. 2.10).

Unsupervised Machine Learning

In **Unsupervised Learning** algorithms, instances are unlabeled [11]. Unlabeled data refers to data that has not been classified in any way. Unsupervised learning algorithms take a set of historical data that contains only inputs and find structures in the historical data. Unsupervised Learning algorithms learn from test data that has not been labeled, classified or categorized. Instead of responding to feedback, unsupervised learning algorithms identify commonalities in the data and respond based on the presence or absence of such commonalities in each new piece of data.

Association is an unsupervised learning approach used to discover the probability of the co-occurrence of items in a collection. It is extensively used for example in market-basket analysis. For example, an association model might be used to discover that if a customer purchases bread, s/he is 80% likely to also purchase eggs.

Clustering is an unsupervised learning approach used to group samples such that objects within the same cluster are similar to each other than to the objects from another cluster.

Association Rule Mining

Association rule mining is a machine learning model that discovers interesting relationships among variables that contain large volume of historical transactional data [1]. A classic use of Association Rule Mining is to analyze transactional data in a point-of-sale process in supermarkets. The idea is to analyze the patterns of purchases of customers and generate rules of frequent buying observations. Association Rule Mining assumes all data is categorical and was initially used for market basket analysis to find how items purchased by customers are related. Market basket analysis derive rules like a customer that buys ham, sliced cheese, and tomatoes, is likely to buy bread { ham, sliced cheese, tomatoes } → { bread }.

This rule is very useful to supermarkets for marketing strategies such as promotional pricing or product placements in the store. Two measures of significance are relevant in Association Rule Mining, these include **Support** and **Confidence** and they are defined below.

Let us assume an example where we have the following tuple observations as rules (where the notation $A.B$ means B follows A):

$$\{ A.B - A.B - A.B - J.B - J.B - J.B - M.N - M.N - M.N \}$$

The following definitions apply:

Let X be an item set, and $X.Y$ a rule, and T the number of total observations in a given dataset.

The **Support** value of X (*supp* (X)) with respect to the set of observations T is defined as the proportion of the observations in the dataset that contain the item-set X. Applying this concept on the above sample dataset:

$$supp(A.B) = \#(A.B) / Tot$$

$$supp(A.B) = 3/9 = 1/3$$

The **Confidence** value of the X.Y rule (conf (X.Y), with respect to the set of observations T is defined as the proportion of the observations in the dataset that contain the item-set X.Y over the number of observations of X in T. Applying this concept on the sample dataset:

$$\text{conf}(A.B) = \# A.B / \# A$$

$$\text{conf}(A.B) = 3 / 3 = 1$$

A more practical application describes these measures of significance in a supermarket situation with the following purchase transactions or baskets that a certain type of customer buys:

Basket 1 = { Juice, Bread, Milk }
Basket 2 = { Juice, Ice-cream }
Basket 3 = { Ice-cream, Beer }
Basket 4 = { Juice, Bread, Ice-cream }
Basket 5 = { Juice, Bread, Fish, Ice-cream, Milk }
Basket 6 = { Bread, Fish, Milk }
Basket 7 = { Bread, Milk, Fish }
Support (observation set) = (# observation set occurs)/total # observations
Confidence (observation set) = (# observation set occurs) / # LHS observation occurs)
If we assume a minimum support minsup = 30% and a minimum confidence minconf = 80%.
Support (Bread, Milk, Fish) = 3/7 = 0.43
Confidence Bread ➔ Milk, Fish = 3/5 = 0.6
Confidence Milk ➔ Bread, Fish = 3/4 = 0.75
Confidence Fish ➔ Bread, Milk = 3/3 = 1
Confidence Bread, Fish ➔ Milk = 3/3 = 1

Sequence Mining

Association Rule Mining does not consider the order of transactions. Sequence Mining, however, takes into consideration the order of transactions in a basket. In Market Basket Analysis, it is interesting to know whether people buy some items in sequence. For example, buying the dinner table first and then buying tablecloth and napkins.

A sequence S is a set $\langle a_1 a_2 \ldots a_r \rangle$, where a_i is an itemset, which is also called an **element** of S. An itemset a_i contains a sequence of elements $\{x_1, x_2, \ldots, x_k\}$. The **size** of a sequence is the number of elements (or item sets) in the sequence; i. e. for the sequence $S = \langle a_1, a_2, a_3 \rangle$ its size is 3. The **length** of a sequence is the number of items in the sequence; i. e. for the itemset sequence $a_3 = (x_1, x_2, x_3, x_4)$ its length is 4.

Measures of significance in an analytic model are important to prioritize or rank the results of the analytic model. In the Sequence Analytic model, two measures of significance allow to rank sequences in order of importance and these are *Support* and *Togetherness*.

Let us assume an example where we have the following tuple sequence observations as rules (where the notation $A.B$ means B follows A):

{ A.B – A.B – A.B – J.B – J.B – J. B – M.N – M.N – M.N }

There are two significance measures that apply to sequence mining and they are Support and Togetherness. These measures of significance are explained below.

Let X be an item set, and $X.Y$ a rule, and T the number of total observations in a given dataset.

The **Support** value of X (*supp* (X)) with respect to the set of observations T is defined as the proportion of the observations in the dataset that contain the item-set X. Applying this concept on the above sample dataset:

supp (A.B) = # (A.B) / Tot
supp (A.B) = 3/9 = 1/3

The **Togetherness** value of the $X.Y$ rule (*tog* ($X.Y$), with respect to the set of observations T is defined as the proportion of the observations in the dataset that contain the item-set $X.Y$ over the number of observations of X or Y in T. Applying this concept on the sample dataset:

tog (A.B) = # A.B / #A/B
tog (A.B) = 3 / 6 = 1 / 2

K-Means Clustering

K-Means Clustering helps to categorize data and to identify categories in your data. It is a great algorithm to identifies non-obvious categories in data. Let us assume we have a set of data point as shown in Fig. 2.11. The observations are not clustered, and the objective is to identify a set of groups in which these data points can be clustered. This can be achieved with K-Means Clustering ML Algorithm.

Fig. 2.11 Initial Centroids Selected by K-Means Clustering Algorithm

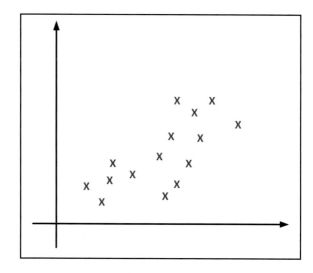

Notice that the algorithm can find clusters not only in two-dimensional data arrays but "n" data arrays. Several steps are followed to create the clusters using the K-Means algorithm.

1. Sep 1 – select the number of clusters "K".
2. Select at random "K" points, the centroids, not necessarily from the dataset.
3. Assign each data point to the closest centroid, and this forms "K" number of clusters.
4. Compute and place a new centroid for each cluster.
5. Reassign each data point to the new closest centroid. If any reassignment took place then go to Step 4, otherwise go to END

The analyst needs to select the number of clusters that he/she desires. There is a mathematical way to identify what is the optimal number of clusters but for now let us consider that we decided to select k=2. Visually, it is not simple as the reader can see that you can select several ways to create these two clusters.

In the second step, the algorithm then identifies the so called "centroids" which are data points associated with the number of clusters. In this case we will have two centroids. The algorithm can select either two points within the dataset that can be used as centroids or the algorithm can select two new random points. Figure 2.12 shows the two centroids selected. The centroids are represented by the blue and red squares in Fig. 2.12. In the third step, the algorithm assigns each data point in the data set to the closest centroid and that will form the k=2 clusters. The two perpendicular lines are used by the algorithm to quickly identify the data points closest to the centroids. Hence, the points above the line are closest to the blue centroid and the points below the line are closets to the red centroid.

The algorithm can use different types of distances and in this case the example uses Euclidean distances. The algorithm then creates two clusters around the k=2 centroids.

Fig. 2.12 Un-clustered set of datapoints

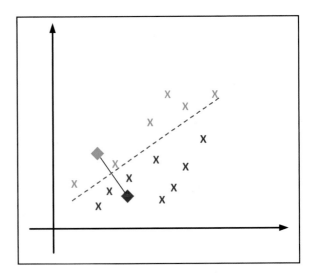

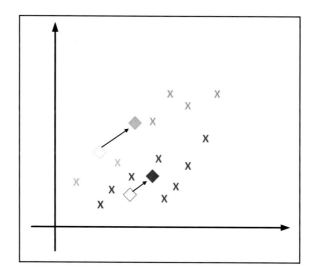

Fig. 2.13 Initial Centroids Selected by K-Means Clustering Algorithm

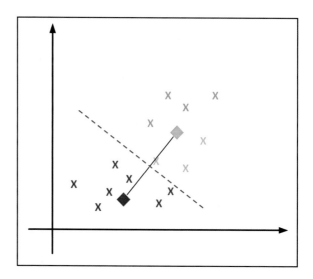

Fig. 2.14 Computation of new clusters

In step 4, the algorithm will compute new values for the two centroids by placing each centroid in the "middle" or center of mass within the cluster. Figure 2.13 shows the location of the new centroids. In the fifth step, the algorithm re-assigns data points to the closest centroid and generates the k=2 new clusters.

The algorithm continues to loop back to step three until it reached stability and points are no reassigned when generating new centroids. Figure 2.14 shows the new reassignment of data points to the computed centroids after the second iteration.

As the iterations continue, the algorithm converges as shown in Fig. 2.15 to form two "stable" new clusters. These then will be the final clusters with their assigned datapoints.

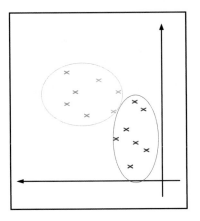

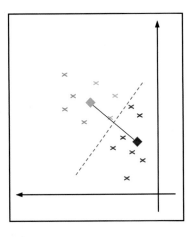

Fig. 2.15 Final convergence of K-means into two stable clusters

Fig. 2.16 Anomaly data points identified by LOF ML algorithm

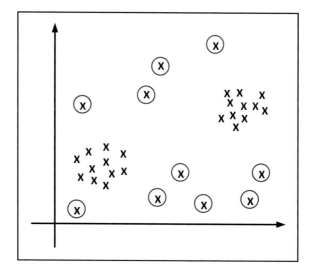

Anomaly Detection Machine Learning

Anomalies are defined as data points in a dataset that appear to markedly deviate from expected outputs. Anomaly detection is the process of identifying data points in a dataset that do not conform to a prior expected behavior in a system. Anomaly detection is being used more and more in the presence of big data that is captured from instrumented objects in an IIoT environment.

Anomaly detection can be conducted in three different ways. First, anomaly detection can be conducted using a Graphical approach by using boxplots, scatter plots, adjusted quartile plots, and symbol plots.

Second, anomaly detection can be conducted using Statistical approaches that include Hypothesis Testing and Scores.

Third, anomaly detection can be conducted using Machine Learning methods. One ML algorithm that is used for anomaly detection is called Local Outlier Factor (LOF) and it is an unsupervised algorithm that computes local densities deviations of a given data point in relationship to its neighbors. It considers as outliers the samples that have a substantially lower density than their neighbors. Figure 2.16 depicts an example of anomalous data points that can be identified by the LOF ML algorithm from a dataset with two clear clusters of data points.

Analytic Conduits

Analytic Conduits in the IIoT are the conduits that are used to analyze data in an IIoT ecosystem. A conduit is a pipeline with specific technology selections that can store data and execute analytic procedures. Analytics conduits are composable in serial and parallel combinations [8]. Conduits provide capabilities to perform data filtering and cleansing, data pre-processing, statistical analyses, data mining analyses, and machine learning.

An essential component of an IIoT architecture is, as shown in Fig. 1.3, the analytics layer. This layer, depending on the types of analytics that need to be conducted, needs to meet the quality attributes required. Quality attributes include performance, reliability, scalability, among others. Depending on where the analytics is performed in the IIoT system, the analytics layer needs to provide a certain quality attribute. For example, if it is needed to conduct analytics at the edge (on a piece of equipment on the shop floor or in the field), the analytics needs to provide results in near-real-time or real-time. While, if the analytics is performed on the Cloud, typically the analytic results do not need to be real-time and hence performance may not be as critical.

When architecting an IIoT system is important to ensure the system is sustainable. Dyllick and Hockerts [5] discuss the importance of architecture sustainability of software systems. The authors identify three elements as key to make software-intensive systems sustainable and these include: (a) technical sustainability; (b) organizational sustainability; (c) financial sustainability.

Technical sustainability in a software-intensive system is achieved by selecting a technology that provides the required quality attributes (or qualities) and provides a platform for future system maintainability and evolution of long-lived systems. Issues such as developers' skills and compatibility with other company's systems are important factors to consider when selecting the appropriate technology. Organizational sustainability refers to having the right resources (people and tools) to ensure development and maintenance are conducted in the most efficient way. Financial sustainability refers to ensuring that the organization meets its expected revenues from the software system. An important element in the financial sustainability is to ensure that the right processes are implemented and followed to reduce non-value-added costs such as re-work, cost of poor quality, etc.

From the technical sustainability perspective, an IIoT architecture must incorporate the non-functional requirements or quality attributes required. An important element of the architecture includes the analytics engine, which, as we discussed earlier, it is the engine of the IIoT system. Given the variety of choices for realizing analytics capabilities, it is difficult and cumbersome for highly specialized domain experts to select the best analytics engine that optimizes the entire system. In software engineering, quality attributes of a software system refer to constraints how the system implements and delivers its functionality [2]. From a Requirements Engineering perspective, functional requirements define operations that the system must be able to perform. Non-functional requirements (NFR) describe how well the system must perform its functions, and how to measure these aspects of the system. Quality attributes are characteristics of NFRs, such as performance, reliability, scalability, security, usability, etc.

It is well understood in the Software Engineering discipline that improving one quality attribute of the system can have negative effects on other quality attributes. In other words, tradeoffs must be made. For example, making a system more secure can make it harder to use, or making it easier to use can make it less secure (security vs. usability), or making it more secure may reduce the performance. Using platform-specific features can make a system run faster, but that often makes it much more costly to port to another platform (performance vs. portability). Trade-offs must be made by taking into consideration the relative importance of each of the system qualities. Many of the important properties of analytics alternatives are quality attributes. This suggests the potential for a systematic approach for choosing the appropriate combination of analytics tools and algorithms for an application domain.

Successful analytics applications in an IIoT environment typically address a problem to solve and use specific technology choices within the IIoT platform. Another application may need to start from scratch, or the previous solution may be reused, even if the trade-offs are ill-suited to the new problem. A core platform provides the common functionality and infrastructure for industrial analytics applications, expediting application delivery. This lowers the knowledge and effort necessary to create business value from machine data. An IIoT core platform needs to be flexible enough to deliver different quality attribute trade-offs when it comes to the choices for industrial analytics applications. Hence, an analytics engine provides a flexible way of performing analytics with our core IIoT platform and needs to be adjusted to address a problem. An analytics engine is an environment with specific technology choices that can store data and execute analytics programs.

Figure 2.17 shows the components and connectors in an Analytics Pipeline. First, the ***Ingestion*** component is connected to the ***Transmitter*** component and has the responsibility to transform the incoming data stream into something the ***Landing Zone Repository*** component can handle. The Landing Zone Repository makes the data available to other components. The ***Analytics*** component is configured with the

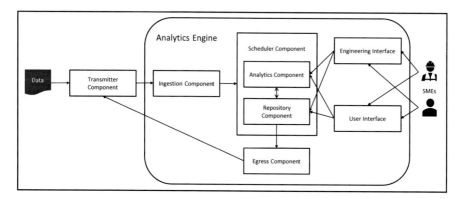

Fig. 2.17 Components of analytics engines

algorithmic functionality, which is designed using the **Engineering Interface**. A **Scheduler** component manages concurrent access to storage and schedules the analytics tasks. Users utilize the **User Interface** to access the data and visualize and use the analytics results. The **Egress** component is responsible for exposing the pipeline results to the system [8].

Analytic conduits have a variety of analytic qualities that need to be considered when designing an analytic application in an IIoT environment. Qualities of analytic engines may include: (a) data flexibility; (b) algorithm flexibility; (c) productivity; (d) static capacity; (e) dynamic capacity; (g) analytics latency; (h) round trip response; (i) scalability; and (j) reliability.

These quality attributes are described in Table 2.1 and they may not be limited to the ones described.

Typical analytic conduit types include Hadoop conduit type, Indexed conduit type, RDBMS conduit type, Streaming conduit type, In-memory type conduit, Single-node type conduit, Graph type conduit, and custom type conduit. Figure 2.18 shows a radar diagram that can be used to evaluate analytic engines quality attributes for specific IIoT applications

Table 2.1 Quality attributes

Property	Description	Scale
Data flexibility	New/unknown data types, without data model modification	1: one data type, 4: limited number of data types, 6: extensible data types, 10: any data
Algorithm flexibility	Variety of supporting libraries, query representations	1: single algorithm, 10: support multiple algorithms
Productivity	Ratio between effort and cost	1: low productivity, 10: high productivity
Static capacity	Store or configure permanently	1:no permanent data storage, 3: GB, 4: TB, 6: PB, 7: EB, 8: ZB, 9: YB, 10: unlimited
Dynamic capacity	Process or manage data simultaneously with concurrent tasks	1:no permanent data storage 3: GB, 4: TB, 6: PB, 7: EB, 8: ZB, 9: YB, 10: unlimited
Analytics latency	Time delay experienced in core data processing	1: decade, 2: year, 3: month, 4: week, 6: day, 7: hour, 8: minute, 9: second, 10: ms
Round trip response	Time elapsed between request and response	1: year, 2: month, 3: week, 4: day, 5: hour, 6: minute, 7: second, 9: ms, 10: μs
Scalability	Ease, speed and affordability of changing performance qualities	1: no scaling, 2: fix number, 4: exponential, 7: limited linear, 10: linear
Reliability	MTBF, operation when faults occur, degree of recovery, no data loss	1: <day, 2: day, 3: week, 4: month, 5: year, 6: decade, 8: century, 9: millennium, 10: > millennium

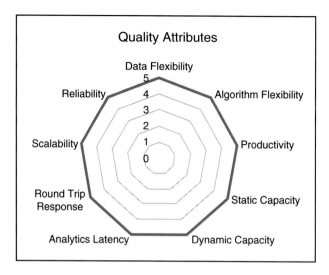

Fig. 2.18 Quality attributes evaluation radar for analytic conduits

References

1. Agrawal, R., Imielinski, T., & Swami, A. (1993). Mining association rules between sets of items in large databases. *Proceedings of the 1993 ACM SIGMOD international conference on management of data*, 207–216.
2. Bass, L., Clements, P., & Kazman, R. (2007). *Software architecture in practice* (2nd ed.). Addison-Wesley.
3. Bishop, C. M. (2006). *Pattern recognition and machine learning*. New York: Springer.
4. Divya, E. (2014). Real time sentiment classification using unsupervised reviews. *International Journal of Scientific and Engineering Research, 5*(3, March), 61–65.
5. Dyllick, T., & Hockerts, K. (2002). Beyond the business case for corporate sustainability. *Business Strategy and the Environment, 11*, 130–141.
6. Gilchrist, A. (2016). *Industry 4.0: The industrial internet of things*. Apress Editors.
7. Glorot, X., Bordes, A., & Bengio, Y. (2001). Deep sparse rectifier neural networks. *Proceedings of the 14th Conference on Artificial Intelligence and Statistics, Fort Lauderdale, FL, USA, 15*, 315–323.
8. Harper, E., Zheng, J., Jacobs, S., & Dagnino, A. (2015), Industrial analytics pipelines. *IEEE big data service conference*, San Francisco, CA, March 30 to April 2nd.
9. Hawkins, D. M. (2004). The problem of overfitting. *Journal of Chemical Information and Modelling, 44*(1), 1–12.
10. Ho, T. K. (1995). Random decision forests. In *Proceedings of the third international conference on document analysis recognition*, Montreal, QC, August 14–16, pp. 278–282.
11. Ho, T. K. (1998). The random subspace method for constructing decision forests. *IEEE Transactions on Pattern Analysis and Machine Intelligence, 20*, 832–844.

12. Kotsiantis, S. B. (2007). Supervised machine learning: A review of classification techniques. *Informatica, 31*, 249–268.
13. LeCunn, Y., Bottou, L., Orr, G. B., & Muller, K. R. (1998). Efficient back-prop. In *Proceeding neural networks: Tricks of the trade, this book is an outgrowth of a 1996 NIPS workshop, ages* (pp. 9–50). London: Springer.
14. Rokach, L. (2010). Ensemble best classifiers. *Artificial Intelligence Review, 33*(1-2), 1–39.
15. https://www.slideshare.net/AngieTabone/sentiment-analysis-78199104
16. https://www.jeremyjordan.me/neural-networks-activation-functions/

Chapter 3
Machine Learning to Predict Fault Events in Power Distribution Systems

The electricity required to supply electric power to cities, is produced in generating stations, or power plants. Such generating stations can be considered as conversion facilities in which the heat and movement energy that originates from fuel (coal, oil, gas, or uranium), sun, wind, or hydraulic energy of falling water is converted to electricity. The Transmission System transports high voltage electricity in large quantities from generating stations to consumption areas. Electric power delivered by transmission circuits must be "stepped down" from high voltage to low voltage in facilities called substations to voltages more suitable for use in industrial and residential areas. The part of the electric power system that takes power from a bulk-power substation to consumers, commonly about 35% of the total plant investment, is called Power Distribution System. Power substations and their equipment play an essential role in the distribution of electricity. Much of the power distribution infrastructure in the US and other parts of the western world is over 50 years old. A key issue facing utilities is to efficiently utilize their limited funds for maintenance and repair of distribution lines. Studies in the UK show that more than 70% of unplanned customer minutes loss of electrical power are due to problems in the distribution grid [9].

Problem Statement

The objective or business opportunity from the IIoT point of view is to forecast fault events in a power distribution grid utilizing historical data on weather conditions, power distribution grid electric value readings at the time of a fault event, and the type of grid infrastructure as shown in the grayed areas in Fig. 3.1. In order to reduce the power outages in the Power distribution grid, it is important to have prediction models that can foretell when a fault event may occur in a distribution network given certain conditions [11]. Power distribution grid operators rely on manual methods and reactive approaches to outage diagnostics and location identification.

© Springer Nature Switzerland AG 2021
A. Dagnino, *Data Analytics in the Era of the Industrial Internet of Things*,
https://doi.org/10.1007/978-3-030-63139-0_3

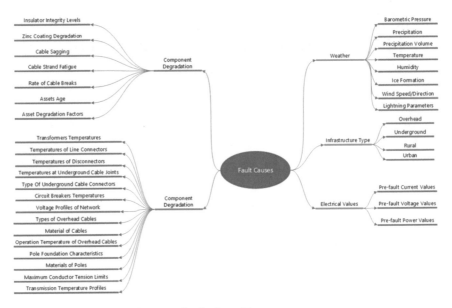

Fig. 3.1 Fault event factors in power distribution grids

Moreover, the efficiency of dispatching crews can be improved by having more automated diagnostic methods and fault predictive capabilities. According to a survey conducted by the Lawrence Berkeley National Laboratory, power outages or interruptions cost the United States of America $80 billion annually [8].

Work in the area of substation and power distribution system fault identification has mostly been reactive, i.e. faults are identified and diagnosed after they have occurred. Fault diagnostics in substations and distribution systems has had limited automation capabilities. Since substation faults are likely to result in costly power outages, forecasting or projecting potential power system fault events before they occur can reduce response time, increase precision, and enhance preparedness to fix the outage, and all these reduce outage costs to both utilities and customers. The objective is to forecast or project fault events and their location in a substation and power distribution grid. Historical fault event and electrical values data of a power distribution system, historical weather data, and infrastructure type data were utilized to create the forecast models. A power utility that implements IIoT concepts is capable of being more proactive in fault forecasting, projection, or prediction.

Background

There are many factors identified in the literature that can cause fault events in a power distribution grid [1, 2, 4, 7, 10, 13, 15, 16]. These factors can be broadly classified into (a) physical properties of the distribution grid; (b) electrical values of grid; (c) weather conditions; (d) assets or components degradation in the grid; and

(e) type of grid infrastructure. A mind-map of the factors associated with fault events in distribution networks is shown in Fig. 3.1.

As mentioned earlier on, the problem statement or opportunity are to forecast potential fault events in a distribution grid using historical data on "normal" weather conditions, grid electric value readings at the time of a fault event, and the type of distribution power grid infrastructure [6, 12]. It is intuitive that weather conditions can have an influence on the distribution grid, especially extreme weather conditions such as tornadoes, hurricanes, and large storms. Nevertheless, even "relatively normal" weather conditions can also have an impact on faults in the distribution grid. Lu et al. [13] discuss the influence that changing climatic conditions and weather have on the wear of electric equipment. Their analyses indicate that during the hot summer months, when the load on a feeder increases to 60–70% there is an over-charge in the lines and assets that can result in increased number of faults and reduced voltage of the energy distributed. Analyses in changes in load consumption across different days of the week based on time of the year and temperature can be conducted. Heine et al. [10] identify several fault events that manifest frequently as a result of lightning storms.

Data for Forecasting Fault Events in Power Distribution Grids

The analytic application presented was developed studying a Power Utility in the US that will be referred to as Investor Owned Utility (IOU). The data collected originate from sensors and equipment of this IOU. The historical dataset types utilized in this work are shown in Fig. 3.2. These datasets include: (a) fault data and electrical values from the IOU feeder system; (b) weather data; (c) infrastructure type of the IOU. The fault data associated with the IOU were collected utilizing sensors and an automated system which consists of intelligent electronic devices (IED's) with sensing and analytic capabilities located at the physical end of the feeders of the distribution lines of the IOU power distribution grid. These IED's monitor electrical values from the distribution lines and can detect a fault event in the grid after it occurred. The fault data includes these electrical values and was corroborated with data entries documented by IOU engineers after restoring service following a fault has occurred. The corroborated fault data accumulates as time passes and becomes an important corpus of fault knowledge used in the analyses. This process is the perfect example of collaboration between sensing technologies and human domain expertise. The weather data were collected from the US National Weather Service (NWS) and from an organization called the WeatherBug (WBUG) weather services. The NWS data were collected by their weather station every five minutes in METAR format. Typical METAR readings contain weather data on temperature, dew point, wind direction, wind speed, precipitation, cloud cover cloud heights, visibility, and barometric pressure. The WeatherBug data was collected from small weather stations located in a variety of schools close to the different

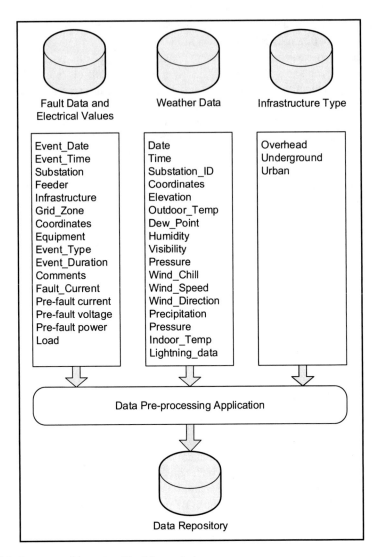

Fig. 3.2 Summary of datasets utilized for analysis

substations of the IOU. Lightning data were also obtained from the WeatherBug weather services organization.

An essential aspect of any data analytics activity is the preparation of the data used for the analyses or "data cleansing and pre-processing" [6]. A software module can automate the pre-processing of the raw datasets, to generate the final dataset that can be used for data analysis. A data pre-processing module can contain rules that address: (a) weather-fault time zone alignments; (b) weather-fault distances; (c) weather direction; and (d) information in free-format comments in the fault events, among others. A brief description of the data pre-processing activities for this study is presented below. Data pre-processing involved cleaning of the fault, weather, and lightning data; aligning all the datasets based on both time and geographic location; and fusing the various datasets together. A total of 1725 fault events were obtained over a 2-year period, across eight feeders in four substations (two feeders per substation) which ranged from a few miles apart to 10 miles apart. The main tasks in cleaning and pre-processing this fault data included generation of mineable parameters from the fault comment texts which described the equipment involved in the fault and the type of problem (e.g. animal contact).

The fault infrastructure coding in the fault events was supplemented with information on the power grid topology, e.g. which feeders were almost entirely underground or overhead. Five-minute weather observation data was obtained from US government sources for one airport within 9–25 miles of the four substations, with 93% completeness (about 100,000 observation records per year). Hourly weather observation data was obtained from WeatherBug[1] for four local "Earth Networks" weather stations within varying distances from the four substations (about 8000 observation records per year for a single weather station). Preprocessing the five-minute airport weather data required several transformation steps: first, to translate the coded "METAR" strings to their equivalent text and numeric parameters; then, to parse and group the weather condition comments into mineable nominal parameters. The hourly WeatherBug data contained somewhat different parameters than the five-minute METAR data (e.g., it included sunlight level readings which the airport data did not contain, and it did not have weather condition comments), but the pre-processing was otherwise similar. We also obtained lightning stroke data for 2009 and 2010 from WeatherBug Total Lightning Network (WTLN)[2] within a five-mile radius of each substation. Since this lightning stroke data was time-stamped to the micro-second, preprocessing it for our data mining purposes required various aggregate counts and sums of strokes and amplitudes for both intra-cloud and cloud-to-ground lightning. These aggregates were calculated for both five-minute periods and one-hour periods so the lightning counts could be joined to the weather data.

To enable selection of the "closest" weather data for each fault, preparing all this data for mining required aligning both the timestamps and the geographic locations. Calculating geographic location parameters was also different for each dataset. For the airport weather data, the precise (lat, lon) of the airport weather station was known. For the hourly weather data, the (lat, lon) values were not available for the weather stations, so the geographic center of each weather station's zip code was used. Since each microsecond-time-stamped lightning stroke had its own unique (lat, lon), a reference (lat, lon) was determined algorithmically to tag the five-minute and one-hour aggregate records. The precise (lat, lon) had not been recorded for many faults, so the (lat, lon) of the associated substation was used for all fault records to ensure consistency. Each fault and weather dataset had a different

---

[1] WeatherBug, http://www.weatherbug.com

[2] WTLN, the WeatherBug Total Lightning Network (http://weather.weatherbug.com/weatherbug-professional/products/total-lightning-network)

reference time zone: some used local time and some used UTC, and some that used local time reflected Daylight Savings while others did not. Therefore, our preprocessing included calculation of a new timestamp parameter for each dataset, adjusted to the same reference time zone (local standard time was chosen). The lightning aggregates were then joined to the five-minute weather data and the hourly weather data. To further prepare for data fusion, approximate 'Great Circle' distances were calculated between each weather station and each substation, and for each lightning aggregate, the distances to the four substations were calculated. These distances were used to choose for each fault the "closest" five-minute and "closest" one-hour weather record, using both timestamp and distance.

Forecasting Fault Events

The first type of analysis that can be performed on the dataset is a set of statistical tests that allow the analyst to profile the data. Figure 3.3 shows an example of the type of statistical analyses that can be performed on such dataset. For example, it is interesting to observe the distribution of electrical faults in the power grid analyzed throughout the day for all years of available data as shown in Fig. 3.3. The graph shows that 6:00–8:00 am and then 13:00–18:00 are two periods where there is an increase of faults during the day. Other interesting statistics can include for example distribution of all types of faults over time, distribution of different types of faults, fault distribution and power demand profiles, correlation between faults and different weather conditions, and others.

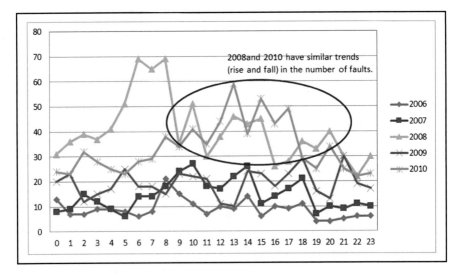

Fig. 3.3 Distribution of all fault types across different hours of the day for years 2006–2010

Several approaches to predict fault events in a distribution grid have been proposed [3, 5, 14]. Butler [2] discusses a failure detection system, which makes use of electrical property parameters (such as feeder voltages, phase currents, transformer windings' temperatures, and noise from the transformer during its operation) to identify failures. Gulachenski et al. [7] use a Bayesian technique to predict failures in transformers [Gulachenski1990]. Some of the features considered in these studies include ambient temperature, varying loads on the transformer, and age-to-failure data of transformers. Quiroga et al. [15] search for fault patterns assuming the existence of relationships between events. Their approach considers factors such as values of over-currents of past fault data and the sequence, magnitude, and duration of the voltage drops. Although the above-mentioned approaches are predictive in nature, they do not consider weather properties to predict faults. The hypothesis associated with the work presented in this chapter is that a fault event occurrence is likely to follow a pattern with respect to weather conditions, infrastructure type, and electrical values in the distribution grid at the time of the fault.

The objective of the analyses presented is to develop an analytics approach that uses machine learning that predicts fault events on the distribution grid based on expected weather conditions. The strategy followed has two phases as shown in Fig. 3.4. In the first phase, the historical data collected is used to and train machine learning models that can foretell fault events using weather forecasts. During the first phase, four algorithms can be utilized to perform five generic analyses (fault prediction, zone prediction, substation prediction, infrastructure prediction, and feeder prediction) and use a percentage of the historical data to train the ML models. The performance, accuracy, recall, and *f*-measure values of the resulting models were then compared for each of the five analyses. During the second phase, the best performing algorithm, which is the Neural Networks model, is selected for each of the five analyses to be utilized for future predictions at the IOU.

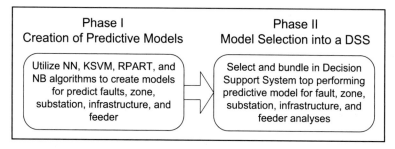

Fig. 3.4 Development phases

Creation of Machine Learning Models

Supervised classification machine learning techniques were utilized to forecast the occurrence of faults in the distribution power grid of the IOU. Four supervised classification machine learning algorithms were utilized to conduct the analyses: Neural Networks (NN), kernel support vector machines (KSVM), decision-tree based classification (recursive partitioning; RPART), and Naïve Bayes (NB). A **NN** is an interconnected group of artificial neurons that use a computational model that allows them to adapt and change their structure based on external or internal information that flows through the network. A support vector machine is a linear binary classification algorithm [15]. Since our datasets consisted of more than two classes, we chose to use kernel support vector machine (**KSVM**), which has been found to work in the case of non-linear classifications. **RPART** is a type of decision tree algorithm that helps identify interesting patterns in data and represents them as a tree. RPART was chosen because it provides a suitable tree-based representation of any interesting patterns that are observed in the data sets, and because it works well with both nominal and continuous data. **NB** is a probabilistic classification technique that is said to work well even with small sets of data. These four algorithms were selected because of their distinct properties and their ability to work with different types and sizes of data, and the objective was to select the algorithm that performed the best for the type of prediction being conducted. Five analyses were conducted utilizing

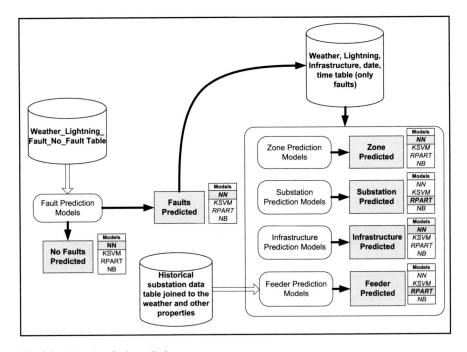

Fig. 3.5 Substation fault prediction strategy

these four algorithms: (a) fault event prediction; (b) grid zone prediction; (c) substation prediction; (d) type of grid infrastructure; (e) feeder number prediction. As mentioned earlier, the primary data attributes shown in the shaded areas of Fig. 3.5 were considered when creating and training the predictive models.

Four models were created to identify weather patterns that are most likely to result in a fault event using the NN, KSVM, RPART, and NM algorithms. The models were constructed by taking weather data points joined to fault events, as well as random weather data samples when no fault events were recorded in the selected IOU substations. Since there were a large number of weather points for the times at which a fault did not occur, a random sample of records was taken making sure that all months and days and hours in the day were covered in the sample. The dataset contained a total of 3471 records (1725 with faults and 1746 without faults), of which 2430 were used for training each of the four models and 1041 for testing the models (see Fig. 3.5).

The output of these models (which corresponds to predictions made on the test data) shows a prediction of the weather conditions for which a fault event may or may not occur (see Table 3.1 for a sample output). The best-performing model was the one created with the feed-forward trained by a multi-layer perceptron back-propagation Neural Network algorithm (see shaded area in Fig. 3.5). This trained model produces an accuracy of 75%, an average precision of 77%, an average recall of 73%, and an f-measure of 75%.

Zone Prediction Models

The four zone prediction models were trained by considering fault historical data from the IOU grid and weather data. Of the 1725 records with faults and weather data, 70% were used for training and 30% for testing the trained models. The output

Table 3.1 Sample output of NN trained fault prediction model

Row No.	ID	confidence(No)	confidence(Yes)	Prediction(Fault)	Outdoor_Te	Humidity%	Pressure(in..	Win_speed..	Wind_direc..	Average_Wi..	Average_Wi..	Lightening...	Out_Temper.
1	3152	1.000	0.000	No	52.265	30.072	29.688	2.411	162	0.438	252	0.000	0.110
2	3153	0.826	0.174	No	32.730	100.000	29.795	1.754	18	3.288	20	0.000	0.110
3	3154	1.000	0.000	No	38.202	79.558	29.906	0.438	0	1.535	320	12.745	0.110
4	3155	1.000	0.000	No	37.637	81.581	29.880	3.508	1	3.288	339	13.523	0.110
5	3156	0.828	0.172	No	30.647	76.978	30.133	1.973	146	1.973	145	0.000	0.110
6	3157	0.536	0.464	No	23.051	77.071	30.111	2.850	171	2.192	172	0.000	0.110
7	3158	0.413	0.587	Yes	31.550	58.203	29.429	0.000	162	0.000	160	0.000	0.110
8	3159	1.000	0.000	No	41.389	24.485	29.563	11.838	275	6.577	269	35.627	0.110
9	3160	1.000	0.000	No	31.887	30.989	29.842	0.438	280	9.865	294	0.000	0.110
10	3161	0.915	0.085	No	41.895	35.824	29.898	2.411	326	4.823	18	49.569	0.110
11	3162	0.181	0.819	Yes	28.675	73.976	29.995	2.411	192	0.577	190	6.980	0.110
12	3163	0.157	0.843	Yes	29.745	100.000	30.198	0.000	206	0.000	206	0.000	0.110
13	3164	0.859	0.141	No	23.447	100.000	29.886	3.069	10	4.384	4	17.967	0.110
14	3165	0.776	0.224	No	17.596	100.000	29.904	3.069	302	3.069	291	0.314	0.110
15	3166	0.986	0.014	No	17.874	99.631	29.902	3.069	29	1.973	46	0.000	0.110
16	3167	0.016	0.984	Yes	22.604	63.484	29.909	5.919	26	5.261	1	69.255	0.110
17	3168	0.645	0.355	No	30.757	100.000	29.957	10.742	287	12.934	310	0.000	0.110
18	3169	0.983	0.017	No	33.337	63.310	30.161	7.892	185	6.138	179	0.216	0.110
19	3170	0.796	0.204	No	32.840	56.174	30.202	7.453	186	5.919	183	0.458	0.110
20	3171	0.916	0.084	No	42.797	26.566	29.966	5.261	181	6.796	176	0.000	0.110
21	3172	1.000	0.000	No	56.970	21.720	29.706	7.234	184	7.892	182	0.000	0.110
22	3173	0.936	0.064	No	43.750	98.863	29.873	4.823	293	5.919	296	0.000	0.110

of these models predicts in what zone (AMZ, UMZ, PMZ) on the IOU grid the fault occurred. The best-performing model was the one created training a Neural Network algorithm as shown in the shaded area in Fig. 3.5. The model contains one hidden layer with 20 nodes. The model produces an accuracy of 66%, an average precision of 69%, an average recall of 68%, and an f-measure of 6.

Substation Prediction Models

The four substation prediction models were trained by considering fault historical data from the IOU grid and weather data. Of the 1725 records with faults and weather data, 70% were used for training and 30% for testing the trained models. The output of these models predicts the IOU substation ID where the fault occurred. The best performing model was the one created with the recursive partitioning algorithm as shown in the shaded area in Fig. 3.5. This trained model produces an accuracy of 59%, an average precision of 66%, an average recall of 54%, and an f-measure of 59%.

Infrastructure Prediction Models

The four infrastructure prediction models were trained by considering fault historical data from the IOU grid and weather data. Of the 1725 records with faults and weather data, 70% were used for training and 30% for testing the trained models. The output of these models predicts the type of infrastructure (overhead or underground) on the section of the IOU grid where the fault occurred. The best-performing model was the one created training a Neural Network algorithm as shown in the shaded area in Fig. 3.5. The model produces an accuracy of 77%, an average precision of 62%, an average recall of 52%, and an f-measure of 57%.

Feeder Prediction Models

The four feeder prediction models were trained by considering fault historical data from the IOU grid and weather data. Of the 1725 records with faults and weather data, 70% used for training and 30% for testing the trained models. The output of these models predicts the IOU Feeder where the fault occurred. The best-performing model is the one created with the recursive partitioning algorithm as shown in the shaded area in Fig. 3.5. This trained model produces an accuracy of 74%, an average precision of 79%, an average recall of 70%, and an f-measure of 74%.

Figure 3.6 displays a graph with the average f-measure values from the four different models created for each analysis. The f-measure is a harmonic mean of the

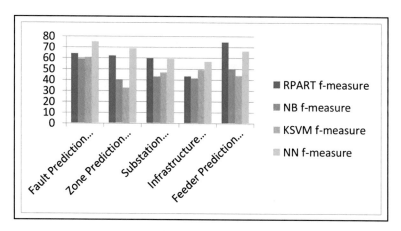

Fig. 3.6 Performance of the different models in terms of their average *f-measure* values

precision and recall of a model and is calculated based on precision and recall using the formula given in Eq. 3.1. Precision pertains to the fraction of classified set of data points that have been correctly classified, and recall is the fraction of the actual set of data points that have been correctly classified. Precision and recall are calculated using the formulas in Eqs. 3.2 and 3.3 respectively.

$$f - \text{measure} = 2 * \frac{precision * recall}{(precision + recall)} \tag{3.1}$$

$$precision = \frac{number\ of\ records\ classified\ correctly\ into\ a\ certain\ class}{total\ number\ of\ records\ classified\ into\ that\ class\ by\ the\ model} \tag{3.2}$$

$$recall = \frac{number\ of\ records\ classified\ correctly\ into\ a\ certain\ class}{total\ number\ of\ records\ that\ actually\ belong\ to\ that\ class} \tag{3.3}$$

Figure 3.7 shows the training and prediction strategy to address the fault prediction in power distribution grids. The diagram shows the training and prediction steps.

Based on the highest *f*-measure of the trained models for each analysis, the following selections can be made for prediction purposes:

(a) NN model is selected to predict faults based on weather;
(b) NN model is selected for prediction of the zone in the grid where a fault may occur;
(c) NN model is selected to predict the substation where a fault may occur;
(d) NN model is selected to predict if the fault occurs in the overhead or underground lines;
(e) Although the RPART model has the highest *f*-measure, the NN model is close and for purposes of consistency it can also be selected to predict the feeder where a fault may occur in the grid.

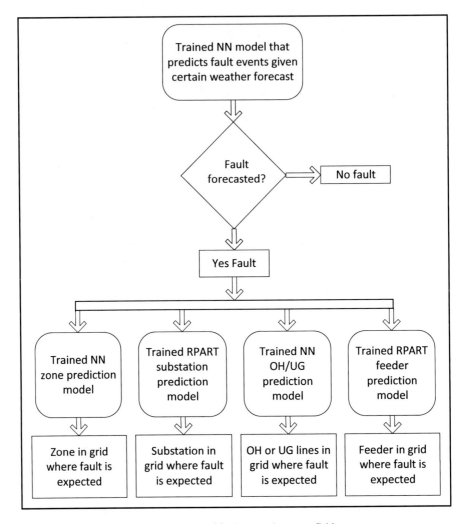

Fig. 3.7 Training and prediction strategy of fault events in power Grid

Proactive Fault Analytics Helps Improving the Business Model and Employee Satisfaction

An advanced analytics application like the one described above and developed and operating at a US utility is an important example of an IIoT application in the power industry. First, the new IIoT approach uses state-of-the-art sensing capabilities deployed at various components in the distribution grid to identify disruptions that occur under "normal" weather conditions. Second, after each disruption, a team that is dispatched to fix the problem, writes a report explaining what the problem was all about. This information is then stored and made available to the analytic system that learns or confirms its knowledge over time to then use it to make recommendations to the power distribution control center. Third, external time-stamped data collected is aligned and incorporated to the dataset and analyzed to create the results used to make recommendations to the users. Fourth, the new approach moves a crucial pain point, such as the fault detection process, from being diagnostic and reactive to being predictive and hence more proactive. Finally, it brings a level of consistency on decision making of power grid Engineers allowing less experienced Engineers to use the system and make advanced decisions in a similar way as their more experienced counterparts.

Utilities, as many industries, face a shortage of domain expertise in many technical areas. Very experienced domain experts are close to retirement age and due to budget cuts in the utility sector, it is difficult to have both experienced SMEs and less experienced Engineers overlap for a long period of time so that there is appropriate technology transfer. While developing a machine learning system that uses historical data and input from SMEs, the system preserves expertise of the domain experts. When less experienced Engineers use the ML system, they learn the associated knowledge that more experienced Engineers possess. This is an efficient way to train new Power Engineers, both in Power Systems technology and computing and AI technologies. The potential for utilities to hire High Qualified Personnel (HQP) to make decisions using the fault prediction analytics system is very high, in the areas of automation and instrumentation. Also, preparing HQP power systems personnel through the knowledge transfer that occurs by interfacing with the IIoT system represents an added benefit to employment.

References

1. Bowers, J. S., Sundaram, A., Benner, C. L., & Russell, B. D. (2008). Outage avoidance through intelligent detection of incipient equipment failures on distribution feeders. In *IEEE power and energy society general meeting – conversion and delivery of electrical energy in the 21st century, 2008* (pp. 1–7).
2. Butler, K. (1996). An expert system based framework for an incipient failure detection and predictive maintenance system. *International Conference on Intelligent Systems Applications to Power Systems, 1996*(1996), 321–326.
3. Chen, W. H., Liu, C. W., & Tsai, M. S. (2002). On-line fault diagnosis of distribution substations using hybrid cause-effect network and fuzzy rule-based method. *IEEE Transactions on Power Delivery, 15*(2), 710–717.
4. Chow, M. Y., & Taylor, L. S. (1995). Analysis and prevention of animal-caused faults in power distribution systems. *IEEE Transactions on Power Delivery, 10*(2), 995–1001.
5. Cortes, C., & Vapnik, V. (1995). Support-vector networks. *Machine Learning, 20*(3), 273–297.
6. Dagnino, A., Smiley, K., & Ramachandran, L (2012, July 1–3). Forecasting fault events in power distribution grids using machine learning. *24th International Conference on Software Engineering and Knowledge Engineering (SEKE'2012)* (pp. 458–463). San Francisco.

7. Gulachenski, E. M., & Bsuner, P. M. (1990). Transformer failure prediction using Bayesian analysis. *IEEE Transactions on Power Systems, 5*(4), 1355–1363.
8. Hamachi, L. K., & Eto, J. (2004, September). Understanding the cost of power interruptions to U.S. electricity consumers.
9. Hampson, J. (2001, November). Urban network development. *Power Engineer Journal, 15*(5), 224–232.
10. Heine, P., Turunen, J., Lehtonen, M., & Oikarinen, A. (2005). Measured faults during lightning storms. In *Proceedings of the IEEE Power Tech 2005*, (pp.1–5). Russia.
11. Lawton, L., Sullivan, M., Van Liere, K., Katz, A., PRS, & Eto, J. (2003). *A framework and review of customer outage costs: Integration and analysis of electric utility outage cost surveys*. Berkeley: Lawrence Berkeley National Laboratory.
12. Lee, H. J., Ahn, B. S., & Park, Y. M. (2002). A fault diagnosis expert system for distribution substations. *IEEE Transactions on Power Delivery, 15*(1), 92–97.
13. Lu, N., Taylor, T., Jiang, W., Jin, C., Correia, J., Leung, L., & Wong, P. C. (2010). Climate change impacts on residential and commercial loads in the Western U.S. grid. *IEEE Transactions on Power Systems, 25*(1), 480–488.
14. Lee, H. J., Park, D. Y., Ahn, B. S., Park, Y. M., Park, J. K., & Venkata, S. S. (2000). A fuzzy expert system for the integrated fault diagnosis. *IEEE Transactions on Power Delivery, 15*(2), 833–838.
15. Quiroga, O., Meléndez, J., & Herraiz, S. (2010). Fault-pattern discovery in sequences of voltage sag events. *14th International Conference on Harmonics and Quality of Power (ICHQP)*, 2010, (pp. 1–6).
16. Yokoyama, S., & Askawa, A. (1989, October). Experimental study of response of power distribution lines to direct lightning hits. *IEEE Transactions on Power Delivery, 4*(4).

Chapter 4
Analyzing Events and Alarms in Control Systems

Process industries use complex control systems to monitor a variety of manufacturing operations. Control systems collect a large variety and volume of sensor data that measure processes and equipment functions. Alarms constitute an integral component of data collected by control systems. These alarms are generated when there is a deviation from normal operating conditions in equipment and processes. With large number of alarms potentially occurring in a plant, it is imperative that operators and plant managers focus on the most important alarms and dismiss unimportant alarms.

Problem Statement

Large and complex processing plants such as chemical facilities, petrochemical industrial, refineries, power generation plants, and similar process industries use complex control systems that help them to keep the processes, equipment and operations working successfully to produce high-quality outputs. Plant operators use control systems that monitor the entire plant in sophisticated control rooms where complex displays are continuously observed. In the early days, control rooms utilized "panel boards" which were loaded with control instruments and indicators that served the function of control systems. These control instruments were coupled to sensors located in the physical plant process streams and on the outside of process equipment [4]. The sensors relayed their information to the control instruments via analogue signals. These systems merely yielded information, and a well-trained operator was required to make the necessary adjustments in the plant based on the information received. Alarms were added in order to alert the operator on a condition that was about to exceed a design limit or had already exceeded a design threshold. Emergency shut down (ESD) systems were also employed to halt a process that was in danger of exceeding either safety, environmental or acceptable process limits. More complex plants had more complex panel boards and more expert human

© Springer Nature Switzerland AG 2021
A. Dagnino, *Data Analytics in the Era of the Industrial Internet of Things*,
https://doi.org/10.1007/978-3-030-63139-0_4

operators. In the early days of panel boards, the number of alarms in a plant was limited by the amount of available panel board space, and the cost of running wiring, and hooking up an annunciator. It was often the case that if you wanted a new alarm, you had to decide which old one you had to give up.

As technology evolved, the control system and control methods continued to advance towards more plant automation. Highly complex material processing called for highly complex control methods. Also, global competition pushed production operations to increase outputs while using less energy and producing less scrap. In the days of the panel boards, a special kind of engineer was required to understand a combination of the electronic equipment associated with process measurement and control, the algorithms necessary to manage the process, and the actual processes that were being utilized to make the products. Around the mid 80's distributed control systems (DCS) began to be used in process industries. A major change during this period was that the engineer could now control the process without having to understand the equipment necessary to perform the control functions. Panel boards were no longer required, because of the information that once came across analog instruments could be digitalized, stored into a computer and manipulated to achieve the same control actions once performed with amplifiers and potentiometers. Alarms became easy and cheap to configure and deploy. You simply typed in a location, a value to set the alarm on, and set it to active. The unintended result was that soon engineers could place alarms in many places. Initial installers set an alarm at 80% and 20% of the operating range of any variable just as a habit. Another consequence of the digital revolution was that what once covered several square yards of real estate of a panel board, now everything had to fit into a few inches of computer monitors. Multiple pages of information were thus employed to replicate the information on the replaced panel board. Alarms were utilized to tell an operator to go look at a page he/she was not viewing. Alarms were used to tell an operator that a tank was filling. Every mistake made in operations usually resulted in a new alarm. Alarms were everywhere. Incidents began to accrue as a combination of too much data collided with too little useful information. It is not just the process industries that require clever alarm management techniques but also other industries like hospitals, where doctors can get increasingly desensitized, immune or overwhelmed by constant hospital medical alarms [20]. Nowadays the trend on Big Data processing, automation, analytics, visualization, cheap sensors, and almost unlimited data storage have continued to increase the sophistication and complexity of control systems. This in turn has increased the need for Alarm Management Systems capable of efficiently simplifying the work of plant operators. Alarm management systems must incorporate ergonomics, instrumentation engineering, systems thinking, advanced analytics, and sophisticated visualization to manage the design of the system to increase its usability. Most often the major usability problem is that there are too many alarms annunciated in a plant upset, commonly referred to as alarms flooding or alarm bursts. There can also be other problems with an alarm system such as poorly designed alarms, improperly set alarm points, ineffective annunciation, unclear alarm messages, and others. It is also desirable to focus operators to as few and important alarms as possible. Another point of interest in modern control systems is to provide a level of prediction before critical events occur and substantive loss in production or even plant equipment occurs. Finally, finding patterns of alarm occurrences become important to predict when bad outcomes could happen.

An important capability that needs to be developed is to have diverse data analytics methods to analyze time stamped alarms in control system to eliminate non-important alarms, optimize the number of alarms shown to operators of process industries, and focus on the most important alarms that need to be followed to avoid costly equipment breakdowns and plant stoppages.

Background

The fundamental purpose of alarm annunciation is to alert the operator about deviations from normal operating conditions of equipment and processes. The ultimate objective is to prevent, or at least drastically minimize, physical and economic loss through operator intervention in response to the condition that was alarmed. For most digital control system users, losses can result from situations that threaten environmental safety, personnel safety, equipment integrity, economy of operation, and product quality control as well as plant throughput. A key factor in operator response effectiveness is the speed and accuracy with which the operator can identify the alarms that require immediate action 4.

By default, the assignment of alarm source points and alarm priorities constitute basic alarm management. Each individual alarm is designed to provide an alert when that process indication deviates from normal. The main problem with basic alarm management is that these features are static. The resultant alarm annunciation does not respond to changes in the mode of operation or the operating conditions. When a major piece of process equipment like a charge pump, compressor, or fired heater shuts down, many alarms become unnecessary. These alarms are no longer independent exceptions from normal operation. They indicate, in that situation, secondary, non-critical effects and no longer provide the operator with important information. Similarly, during startup or shutdown of a process unit, many alarms are not meaningful. This is often the case because the static alarm conditions conflict with the required operating criteria for startup and shutdown. In all cases of major equipment failure, startups, and shutdowns, the operator must search alarm annunciation displays and analyze which alarms are significant. This wastes valuable time when the operator needs to make important operating decisions and take swift action. If a resultant flood of alarms becomes too great for the operator to comprehend, then the basic alarm management system has failed as a system that allows the operator to respond quickly and accurately to the alarms that require immediate action. In such cases, the operator has virtually no chance to minimize, let alone prevent, a significant loss. In short, it is important to extend the objectives of alarm management beyond the basic level. It is not enough to utilize multiple priority levels because priority itself is often dynamic. Likewise, alarm disabling based on unit association or suppressing audible annunciation based on priority do not provide dynamic, selective alarm annunciation. The solution must be a Smart Alarm Management (SAM) system that can dynamically filter alarms based on the current plant operation and conditions so that only the currently significant alarms are annunciated. The fundamental purpose of alarm annunciation is to alert the operator to relevant abnormal operating situations. The ultimate objectives are no different from the previous basic alarm annunciation management objectives. Dynamic alarm annunciation management focuses the operator's attention by eliminating extraneous alarms, providing better recognition of critical problems, and insuring swifter, more accurate operator response. Moreover, dynamic alarm annunciation can provide a level of "projection" of a critical event.

Figure 4.1 shows the levels of maturity or phases that an industrial site can have in its alarm management capabilities 4. There are primarily 5 levels or stages and these include: (a) level 1, when the alarm management process and system is overloaded; (b) level 2 where the industrial site is reactive to alarms; (c) level 3, where the industrial site is stable; (d) level 4, where the industrial site has a robust alarm management process and system; and (e) where the industrial site is predictive.

A modern Distributed Control System (DCS) can be a perfect example of an IIoT system as shown in Figure 4.2. As explained in earlier chapters, the IIoT refers to interconnected sensors, instruments, and other devices networked together with various computers' industrial applications. This inter-connectivity allows for efficient data collection, data exchange, and sophisticated data analysis, that lead to improvements in productivity and efficiency and other economic benefits [5]. The IIoT can be considered an evolution of a DCS that allows for a high degree of automation by using cloud computing to refine and optimize

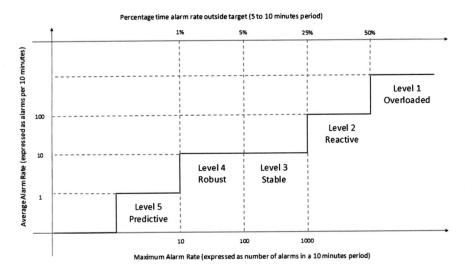

Fig. 4.1 Maturity levels of alarm management (EEMUA 191 – Performance level (Ed 2))

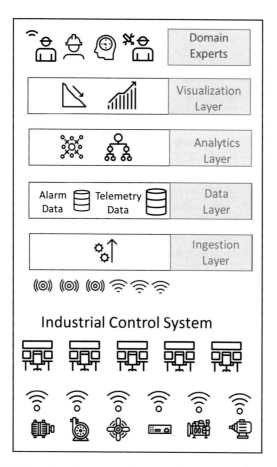

Fig. 4.2 An IIoT DCS including alarm data collection and analysis

process controls. IIoT DCS implemented in process industries such as refineries, petrochemical, oil and gas, and pulp and paper, monitor all production components and processes in the plant. Sensors integrated in the equipment and processes in the plant send a large variety of data such as temperatures, pressure values, concentration values, viscosity values, and in general all telemetry data extracted from instrumentation, and alarm and events data which are time stamped. The ingestion layer in the IIoT DCS gets the data from the plant and stores it into the Data Lake/Data Warehouse in the data layer. Once the data is stored, then proper partition of the data is created into the alarm data which is also pre-processed and prepare it for analysis. The analytics layer consists of statistical and machine learning models that are used to analyze the alarm data. The results of the analyses are presented to the alarm plan alarm subject matter experts to make decisions in the alarm rationalization meetings [16] in the visualization layer utilizing visualizations and GUI tools as shown in Fig. 4.2.

Anatomy of Alarms in IIoT Distributed Control Systems

Process related alarms in an IIoT DCS environment are activated when a sensor value or measurement is out of the accepted threshold. When the abnormal situation is back to acceptable ranges the alarm is de-activated. Fig. 4.3 shows a one-hour time interval with alarms being triggered and returning to normal in a control system. The diagram also shows that an operation action occurred in the log.

Alarms are time stamped events that have an *activation time stamp* (Act) and a *return to normal time stamp* (RTN). The difference between the return to normal time stamp and activation time is the *alarm duration*. Figure 4.3 shows six alarms that are activated and return to normal in an hour period. Notice that in real life, in one hour there are typically many more occurrences than the ones shown. Alarms have also other characteristics that can be captured as part of their data attributes and these include: (a) Alarm ID; (b) Activation Time; (c) Return to Normal time; (e) Alarm Source; (f) Priority; (g) Time Zone; and others. Figure 4.3 also shows that at a point in time an operator action Oper_Action_M also occurs with an associated time stamp. Some definitions at this point are pertinent:

Definition 1 An alarm returns to a normal state RTN before it activates again ACT.

Definition 2 The length of an alarm log is typically selected by the analyst and is measured in time units (T)

Definition 3 The Frequency of an alarm A across the entire data log is denoted as n_A and it is the number of times the alarm activates and returns to normal during T.

Definition 4 The global rate of an alarm A is denoted as GR_A across the entire log T is n_A/T

Definition 5 The local rate of alarm A in a given period of time [ti, tj] within T, is denoted as $LR_{A[i, j]}$ and it is equal to $n_{A[i, j]} / [ti, tj]$

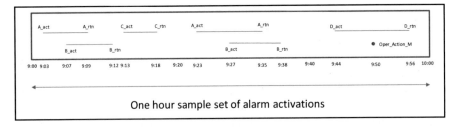

Fig. 4.3 Sample alarm log

Definition 6 The life of alarm A at time t is the time span between the activation time stamp *Act* t_i and its return to normal *Rtn* at time t_{i+j} and is expressed as LA @ t $= (t_{i+j} - t_i)$

Definition 7 The global average life of alarm A GLA is equal to the sum of all lives divided by the number of activations of alarm A in a particular time period.

Definition 8 Let us define G_A as the gap size (in time units) of alarm A between the time it returns to normal Rtn and it is activated again Act, then Ga @ t_i is the gap of alarm A activated at time $t_i = t_i - t_{i-1}$

Definition 9 The global average gap of alarm A is GAG_A is equal to: GAG_A = SUM $[(t_i - t_{i-1})]$ / n, where n is the total number of activations and return to normal of alarm A during the time interval considered.

Two important concepts in alarm analysis are (a) lifetime or life duration of an alarm; (b) time gap between alarms. These concepts are explained below.

The *lifetime* of an alarm is defined as the time between its Rtn, and Act, times. Every time that something occurs in an alarm management system it is recorded with its specific time stamp and stored in the historical event log. An alarm has associated two-time stamps. First, its activation time (Act,) occurs when an alarm is activated due to a threshold violation, operator action etc. Second, its return to normal (Rtn,) occurs when an alarm is tuned off due to an operator action or to a change to a normal threshold level value. The lifetime of alarm A is then expressed as:

$$LT_A = \left[Rtn_{t_A} - Act_{t_A} \right] \tag{4.1}$$

The *time gap* between alarms is simply the amount of time that exists between each activation times of an alarm within the historical alarm log. So, the time gap between alarm A and alarm B given that the activation time of B occurs after the activation time of A is defined as:

$$TG_{B-A} = \left[Act_{t_B} - Act_{t_A} \right] \tag{4.2}$$

Alarm Data

The historical data stored in alarm management systems typically contains alarms, and operator actions with their respective time stamp. Alarms are activated at a particular time stamp and then they return to a normal status at a subsequent point in time. A generic alarm management data model has been developed in this project and is based on the ANSI/ISA 18.2 alarm management standard and it contains 25 data attributes [4]. For purposes of the analyses conducted in this chapter the

Table 4.1 Alarm data attributes and their description

Attribute No	Attribute Name	Details with example
1	TimeStamp	Timestamp when the event occurs. Can have various formats but most used is MM/DD/YYYY hh:mm:ss.msec AM/PM
		e.g. 7/19/2014 6:09:27.527 PM
2	Active Timestamp	Timestamp linking the various alarm status such as RTN, Inactive to its source when it is active
3	Priority	Priority of the alarm typically assigned by an ordinal value. Lower number reflects higher severe situation e.g. 1. This attribute serves to identify if an alarm is critical.
4	Condition	An attribute indicating the associated condition of the alarm such as H, HH, L, LL etc. This attribute in conjunction with "Priority" helps to rank the severity of the alarm and also to identify the name of the alarm
5	Device	Plant device associated with alarm
6	Process Area	Process area in the plant where the alarm originates
7	User	Operator viewing or operating on an alarm typically a descriptive variable indicating operator name or operator workstation e.g. John or OL-PCD\erlst
8	Category	Event Category ID typically an ordinal value such as 666371

following alarm data attributes are needed: (a) TimeStamp (typically activation time for an alarm, "ACT"); (b) Active TimeStamp (typically return to normal status for an alarm, "RTN"); (c) Priority; (d) Condition; (e) Device; (f) Process Area; (g) User; and (h) Category. Table 4.1 summarizes the meaning of each data attribute.

Alarm Management Analytics Models

The objective of alarm management is to reduce the number of alarms an operator views so that the operator can focus her/his attention to those alarms that pose the greatest danger to plant integrity. The analytics example described in this chapter provides results that are analyzed in the so called "Alarm Rationalization" activities. As explained early in this chapter, the design and process Engineers decide based on their domain expertise how to define alarms in the IIoT DCS. As discussed earlier, this often results in a large set of alarms that are triggered and displayed to the operators during the plant operations. Hence, most process industries have periodic "Alarm Rationalization Meetings" to analyze historical alarm occurrences and make decisions on which alarms should be removed (nuisance or redundant alarms) from the IIoT DCS, which alarms should be kept, and which alarm patterns should also be kept in the system and shown to the operators to ensure safe operation of the plant.

The analysis of historical occurrences of alarms described in this chapter using Data Science and Machine Learning approaches, are designed to provide valuable information on alarm patterns and facilitate the Alarm Rationalization Meetings so that domain experts can make decisions on what alarms and alarm patterns should be kept so that operators can focus on important alarms and alarm patterns and they become more effective when they monitor plant operations and potentially floods of alarms.

There are many ways to analyze historical alarm logs such as alarm de-chattering, alarm containment analysis, alarm flood analysis, alarm sequence analysis, and critical event analysis among others. For the purpose of this chapter, the focus will be on *Alarm De-chattering* and *Alarm Sequence Analysis*.

Chattering alarms or nuisance alarms do not have practical value to operators as they do not represent dangerous conditions and keep occurring over. Hence, it is convenient to identify them and once they are considered chattering alarms in the Alarm Rationalization Meetings can be removed from the IIoT DCS. Identify frequent recurring of alarm sequences is very important to identify patterns of historical alarm successions that occur over time. During Alarm Rationalization Meetings, the plant domain experts can then include these sequences in the IIoT DCS so that operators can prognosticate alarm sequences and their time intervals to take actions before a serious high importance alarm or event occur. It is noteworthy to notice that given the importance and priority an alarm can be identified as *Critical* and when this alarm appears in a sequence then the operator will have to monitor the sequence very carefully. Alarm de-chattering and alarm sequence analysis are conducted utilizing data mining and Machine Learning techniques that will be described in detail in this chapter. Association rule Mining, Sequential Data Analytics, and concepts of Market Basket Analysis are used to conduct the alarm analyses. The next section describes these concepts in more detail.

Sequence Pattern Mining and Association Rule Mining

Agrawal and Srikanth [1] define the Sequential Pattern Mining problem as "Given a database of sequences, where each sequence consists of a list of transactions ordered by transaction time and each transaction is a set of items, sequential pattern mining is to discover all sequential patterns with a user-specified minimum support, where the support of a pattern is the number of data-sequences that contain the pattern". Garofalakis [8] describe the sequential pattern mining problem as given a "... set of data sequences, the problem is to discover sub-sequences that are frequent, i. e. the percentage of data sequences containing them exceeds a user-specified minimum support". Masseglia et al. [10] , describe the sequential mining problem as the ..." discovery of temporal relations between facts embedded in a database", while Zaki [18] describe it as a process to "... discover a set of attributes, shared across time among a large number of objects in a given database". Manilla et al. [11] describe the problem as an episode that is defined as a "... collection of events that occur relatively close to each other in a given partial order". Mooney and Roddik [12] provide a comprehensive survey on algorithms in the literature that have been used to address the sequential mining problem. The authors describe different algorithm types. *A priory-based* algorithms are used to discover intra-transaction associations and generate rules about the associations 1. Pattern Growth algorithms were developed to improve performance and are especially useful in very large datasets. Pattern Growth algorithms can be more difficult to develop and maintain than A priori-based algorithms. The FP-Growth algorithm [9] is an example of Pattern Growth models. Temporal Sequence algorithms have been developed to analyze series of events that occur at specific times to determine events that occur frequently together. Manilla et al. [11] developed the first algorithmic framework to mine episodic datasets where they defined an episode as "… a collection of events that occur relatively close to each other in a given partial order" where they defined a frequent episode as "… a recurrent combination of events". Algorithms that mine frequent sequence patterns and item-sets can have performance issues. This has led to the development of algorithms that produce Frequent Closed Item-sets, that produce smaller, yet complete set of results. These algorithms are typically extensions of algorithms that mine the complete set of sequences that improve pruning of search spaces and their traversing [13, 14, 19].

Alarm Baskets

Creation of time bins or time baskets plays an important role when conducting alarm analyses. There are different types of baskets needed for each analysis. The first type of baskets are *Chattering Alarm Baskets* are generated in the De-chattering analysis model to store chattering alarms where the time between their occurrences is very short (short time gap) such as alarm B1, or have short time duration (short life time) such as alarm B2, as shown below in Fig. 4.4.

The second type of baskets are *Equal Time Baskets* that are used to analyze alarm sequences that divide the alarms log into equal time periods. As an example, let us take an alarm log with an hour duration as shown in Fig. 4.3. This event log as explained above shows the activation time stamps and return to normal time stamps of alarms that occur in a control system. The alarm log can be subdivided into equal time slices. The analyst decides the time slice to be used. Figure 4.5 shows that the analyst decided to create 3-time bins or baskets of 20 minutes per slice. Twenty minutes slices are a good size of time basket based on experimental data conducted in this research. It is possible to observe that time slices can "cut" an alarm occurrence in such a way that in one basket an alarm is activated and in a difference time basket the same alarm returns to normal as shown in the case of alarm "C". Equal time baskets are used to conduct deep de-chattering of alarms and sequence alarm mining. When conducting deep de-chattering and of alarm logs, we consider both activation and return to normal conditions of alarms. When conducting sequence mining and critical alarm analyses of alarm logs, we consider only activation times of alarms, and return to normal time stamps are removed from the analysis, as we are interested in sequences of alarms that follow each other from the activation point of view. Figure 4.5 shows at the bottom the alarm time stamped contents of each basket.

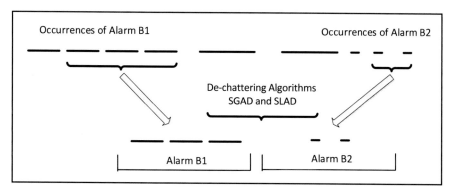

Fig. 4.4 Chattering alarms: Short time gap/short time duration

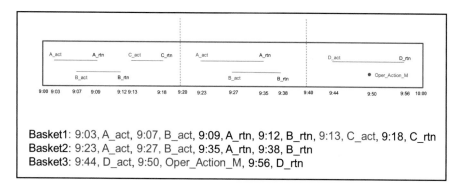

Fig. 4.5 Alarm equal time slices

Alarm De-chattering Analysis

Figure 4.6 shows the different types of chattering alarms where the time interval goes from $t = t_0$ to time $t = t_1$ as shown in the timeline. Chattering alarms are shown as dashes. The size of each dash represents the lifetime or duration (Act, Rtn) of the alarm in relative time , while the space between dashes represent the time gap between the Rtn time and Act time of the same alarm. Utilizing definitions 1 through 9, we define an alarm as a chattering alarm if it has: (a) a short life and a short gap; (b) a normal life and short gap; (c) a short life and a normal gap; a concentrated short life and short gap; (d) chattering burst and then normal.

The primary objective of the de-chattering analysis is to identify chattering alarms in a log and mark them for removal at the Alarm Rationalization meeting. The objective is that once the alarm log is free of chattering or nuisance alarms, then the alarm patterns will have a higher degree of significance. An important thing to notice is that if chattering alarms are left in the alarm event log, these alarms could appear as alarm sequences when doing sequence-mining, but these sequences will not be "interesting".

The algorithms used to identify the chattering alarms are not from the Machine Learning, but they are included in this analysis to show the importance of removing these chattering alarms before a more sophisticated analysis can be done. Two alarm de-chattering algorithms are used, and both utilize the concept of "basketization".

The first algorithm called Short Life Alarm De-chattering (SLAD) focuses on the life span of an alarm and then on the frequency at which an alarm repeats in the alarm log. If an alarm has a short life (it is a parameter defined by the analyst but typically is as having a duration of less than 30 secs) and is repeated several times in a row (three or more times), then the algorithm generates a basket of these short life alarms and keeps the first alarm in the file to be analyzed and marks as chattering alarms all repeated alarms in the basket. Chattering alarms 1, 3, and 4 in Fig. 4.6 are then marked as chattering alarms with potential to be eliminated from the alarm log, leaving only the last occurrence of the alarm in the cleaned alarm log file (depending on the analyst's choice).

The second algorithm is called Short Gap Alarm De-chattering (SGAD) and focuses on repetitive alarms where alarm gaps between the activation and return to normal of an alarm has a short gap (this is defined by the analyst but typically is equal or smaller than 10-20 seconds). From Fig. 4.6, the SGAD algorithm will pick up alarms Similarly, the second algorithm focuses on alarms 1, 2, and 4. These alarms are placed in the chattering alarm basket and are flagged as chattering alarms and if they are not considered as important in the Alarm Rationalization meeting, they can be removed from the IIoT DCS system.

The alarm de-chattering should be conducted before any other analysis. This analysis reveals those alarms which appeared very often and/or with short time life and should be analyzed in more detail. The de-chattering result shows the alarms that are considered as chattering and the analyst has the option of removing them from the alarm log. To get results that match their requirements, the user should make several tests with different de-chattering parameters such as lifetime, time gap, number of repeated alarms, and select if alarm left is the first one or the last one. Figure 4.7 shows the primary steps that the SGAD algorithm follows. For the SLAD de-chattering algorithm, just replace steps 1 and 5 in the SGDA algorithm.

A way to visualize the alarm de-chattering analytics is presented in Fig. 4.8. The visualization is divided into two panes. The left pane contains the names or ID's of

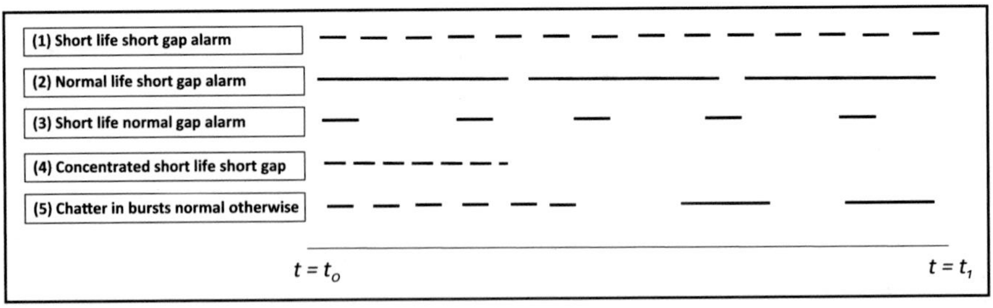

Fig. 4.6 Types of chattering alarms

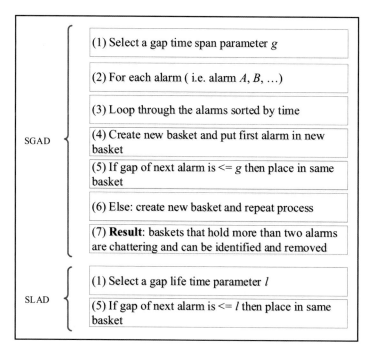

Fig. 4.7 Alarm de-chattering algorithmic process

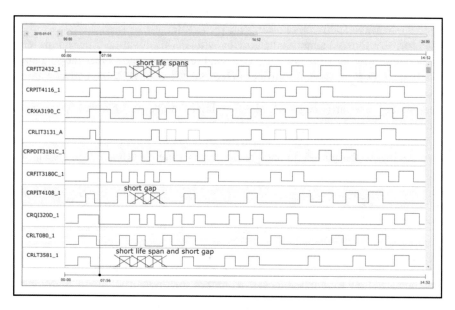

Fig. 4.8 Visualization after alarm de-chattering analytics

the alarms while the right pane shows the timeline and the occurrences of the alarms in this timeline. A flat period means no alarm has been activated or returned to normal. A step means that an alarm has been activated and the length of the step represents the activation time of the alarm and hence its lifetime. When the step goes down that is the point in time when the alarm has returned to normal. A cross in the visualization means that the alarm is chattering due to a certain reason (explained in red) and will be flagged for removal form the alarm log to then perform alarm analytics.

Alarm Sequence Analysis

Once the alarm de-chattering has been completed, during the Alarm Rationalization meeting decisions are made on which alarms are chattering or nuisance and these can be eliminated. With the file of removed nuisance alarms, the analyst has a "clean" alarm log without nuisance alarms and the file is ready to conduct other types of analyses such as the Sequence Alarm Analysis [2, 3, 6]. An important data pre-processing step for Sequence Analysis is not to consider the Rtn time stamps and only consider the Act time stamps. The Alarm Sequence Analysis Algorithm considers a log of historical alarm data (only activation times) and determines the sequences of alarms that frequently occur together. These alarm sequences can be of any length. Sequences of length "two" (2) are also called "parent-child" alarms. The Sequence Analysis model uses concepts of Association Rule Mining and Sequence Mining and calculates the average and standard deviation time between occurrences of each element in the sequence, as well as the minimum and maximum time of occurrence between two elements [3, 7, 15].

Figure 4.9 shows a simplified alarm log with alarm activation time stamps (ts_i), subdivided into four-time intervals $T=R$. Figure 4.9 also shows that the sequence of alarms A → B → C occurs 4 times in the event log and the time between alarm occurrences is calculated as the difference among time stamps of each occurrence. The Sequence Alarm analysis has as primary objective to identify frequent sequences of alarms that occur in an alarm log, the number of times the sequences occur, the average times among alarm occurrences within the sequence, the standard deviation, the minimum and the maximum time occurrences among alarms within the sequence.

Measures of Significance or Metrics for Sequence Analysis

Measures of significance in an analytic model are important to prioritize or rank the results of the analytic model. In the Sequence Analytic model, two measures of significance allow to rank sequences in order of importance and these are **Support**

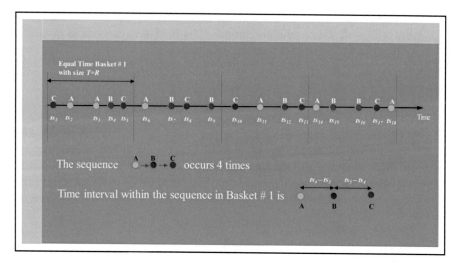

Fig. 4.9 Example of alarm activation sequences

and *Togetherness*. Let us assume an example where we have the following tuple observations as rules (where the notation *A.B* means *B* follows *A*):

$$\{\,A.B - A.B - A.B - J.B - J.B - J.B - M.N - M.N - M.N\,\}$$

The following definitions apply:

Let *X* be an item set, and *X.Y* a rule, and *T* the number of total observations in a given dataset.

The **Support** value of *X* (*supp* (*X*)) with respect to the set of observations *T* is defined as the proportion of the observations in the dataset that contain the item-set *X*. Applying this concept on the above sample dataset:

$$supp(A.B) = \#(A.B) / Tot$$

$$supp(A.B) = 3/9 = 1/3$$

The **Confidence** value of the X.Y rule (conf (X.Y), with respect to the set of observations T is defined as the proportion of the observations in the dataset that contain the item-set X.Y over the number of observations of X in T. Applying this concept on the sample dataset:

$$conf(A.B) = \# A.B / \# A$$

$$conf(A.B) = 3/3 = 1$$

The **Togetherness** value of the $X.Y$ rule (tog ($X.Y$), with respect to the set of observations T is defined as the proportion of the observations in the dataset that contain the item-set $X.Y$ over the number of observations of X or Y in T. Applying this concept on the sample dataset:

$$tog(A.B) = \#A.B / \#A / B$$

$$tog(A.B) = 3/6 = 1/2$$

From the example shown in Fig. 4.9, the Sequence Mining algorithm creates four equal time baskets of length R and in each basket contains the alarms that are recorded in the alarm log and they are ordered according to time stamps (when they were activated). So, there are four equal time baskets Bi for $i = 1, 4$, where the notation $C.A.A.B.C$ means that alarm C follows alarm B, follows alarm A, follows alarm A, follows C.

$$B_1 = \{C.A.A.B.C\}$$

$$B_2 = \{A.B.C.B\}$$

$$B_3 = \{C.A.B.C\}$$

$$B_4 = \{A.B.B.C.A\}$$

The following frequent sequences are then identified as _Sfs_ where _Sfs_ = {A:4; AA: 2; AB: 4; ABB: 2; ABC: 4; AC: 4; B:4; BB: 2; BC: 4; C:4; CA: 3; CAB: 2; CABC: 2; CAC: 2; CB: 3; CBC: 2; CC: 2}. Wang et al. [17] discuss the importance of identifying closed sequences because they lead not only a more compact yet complete result set but also better efficiency in terms of calculation and usage. The complete set of frequent closed sequences S_{fcs} is S_{fcs} = {AA: 2; ABB: 2; ABC:4; CA: 3; CABC: 2; CB: 3} which has only six sequences. Wang et al. 17 identify that S_{fcs} is more compact than S_{fs}. For example, frequent sequence CBC:2 is absorbed by sequence CABC:2.

The sequence mining algorithm identifies closed sequences of alarms as presented above with support supp = 2. To rank the above sets of closed sequences we apply the togetherness parameter and the final ranking becomes: tog = AA = 2/5; ABB = 2/6; ABC = 4/6; CA = 3/6; CABC = 2/6; CB = 3/5. In this case the highest ranked sequence will be ABC with support = 2 and togetherness = 0.67.

A typical visualization of alarms sequence of industrial alarms from a process plant after Sequence Analysis has been conducted looks as shown in (Fig. 4.10).

The alarms are visualized as nodes ad the sequence between nodes as directed arrows that go from the origin alarm towards the destination alarm. The color scheme

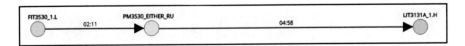

Fig. 4.10 Visualization of an alarm sequence

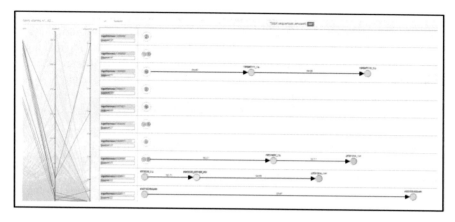

Fig. 4.11 Visualization prioritized alarm sequences

in the node denotes the severity of the alarm (green as less severe, then blue, yellow, orange, and red is a critical alarm). The alarm IDs are shown on top of the nodes. The arrows also show the average observed time between the activation time of one alarm and its subsequent alarm considering the whole alarm log. In this manner, the various sequences of alarms can be shown in a visualization as shown in Fig. 4.11.

Within a sequence a critical alarms or tripping events can be at the end of the sequence and these are states in an IIoT DCS that plant operators and managers want to avoid. All alarms have associated an attribute that determines its severity level in terms of high, medium, and low (LL; LM; LH; MM; H; HH). Sometimes, the plant operators have also a list of critical alarms or events that wish to track during the plant operation and identifying frequent recurring sequences with high priority alarms is very important.

Enhancing Expert Knowledge of Plant Operations Through Advanced Analytics Alarm Management

An advanced analytics alarm management IIoT system implementation as the one described in this chapter provides enormous benefits not only to the process industries that use and adopt this innovative solution but also the IIoT OEM solution provider. Benefits include economic gains, enhanced knowledge level in the organization, contribution to increased safety, enhanced job satisfaction, and in many instances increased job generation. Let us discuss first some of the benefits to the IIoT OEM that provides the advanced analytics solution to its customers.

An OEM that produces an IIoT alarm management solution for its customers will open the possibility of a new business model for itself. The traditional approach would be to offer an advanced analytics solution integrated with the DCS software that customers buy. Using this approach, only very large and "wealthy" process industries can afford such a solution. As an IIoT organization, the OEM can place the alarm management software solution on the Cloud and sell subscriptions to large and small clients, customizing the solution for each application and making it more affordable to all. An IIoT alarm system provides possibilities for new revenue sources, such as pay-per-use revenue models to allow customers to avoid high up-front investments on alarm management software licenses. Another benefit is the capability of OEMs to offer the alarm management solution to other non-traditional industrial customers such as hospitals and health care facilities, computer intrusion-detection systems, and others. This also benefits the OEM from the perspective that developers can learn new domains and new jobs are likely to emerge to continue developing and enhancing the alarm management solution. Another benefit to the IIoT OEM solution provider is the increased job satisfaction to its employees due to the recognition of its customers that the alarm management solution enhances the operations, safety, efficiency, and productivity of their operations. One more advantage is that the IIoT OEM can offer remote alarm management services to their customers and to do this, new high qualified personnel needs to be hired.

Customers from the IIoT OEM are likely to have important benefits from the advanced analytics alarm management solution as it not only helps improve safety of operations in process plants but also helps improve the efficiency of plant operators, increase consistency, focus plant operators on most important alarms that can help avoid critical events, and also increase job satisfaction. An IIoT alarm system brings together technical people in the plant from various domains and fosters collaboration and breakup of silos in the operation. A system like the one described above helps improving the quality of work of plant operators as it reduces the level of stress specially in emergency situations as it allows the operators to focus on the most important alarms. When younger and less experienced plant personnel interact with the intelligent alarm management system, they learn from the system where vulnerabilities in their plant may exist and with this information, they can analyze how to harden those exposed areas.

Members of the alarm rationalization team at the customer site use the results of alarm analytics to identify and address a comprehensive and prioritized list of alarm and event patterns that occur in the plant and develop robust strategies to address emergencies and address areas of weaknesses in the operation. An IIoT alarm management system allows its users to improve their maturity in terms of alarm management by implementing a more holistic approach to the alarm management process. As a summary, the customers that use the IIoT alarm management system achieve a regulatory safety improvement, operator efficiency, favored insurance premiums, increased production, reduced maintenance costs, reduce unplanned downtime, and increased capacity utilization.

References

1. Agrawal, R., & Srikant, R. (1994). Fast algorithms for mining association rules. In J. B. Bocca, M. Jarke, & C. Zaniolo (Eds.), *20th international conference on very large databases (VLDB)* (pp. 487–499). Santiago: Morgan and Kaufmann.
2. Agrawal, R., & Srikant, R. (1995). *Mining sequential patterns. IBM Research Report.*
3. Agrawal, R., Imielinski, T., & Swami, A. (1993, May). Mining association rules between sets of items in large databases. In *Proceedings of the ACM SIGMOD conference on management of data*, Washington, DC.
4. ANSI/ISA-18.2-2009 management of alarm systems for the process industries (2009).
5. Boyes, H., Halaq, B., Cinningham, J., & Watson, T. (2018, October 1). The Industrial Internet of Things (IIoT): An analysis framework. *Computers in Industry, 101*, 1–12.
6. Dagnino, A. (2019). Data mining methods to analyze alarm logs in IoT process control systems. In *2019 IEEE 15th international conference on automation science and engineering (CASE)*, Vancouver, BC, Canada, pp. 323-330. https://doi.org/10.1109/COASE.2019.8843098.
7. Dietterich, T. G., & Michalski, R. S. (1985). Discovering patterns in sequences of events. *Artificial Intelligence, 25*, 187–232.
8. Garofalakis, M. N., Rastogi, R., & Shim, K. (1999). SPIRIT: Sequential pattern mining with regular expression constraints. In *25th international conference on very large databases, VLDB'99*, Edinburgh, Scotland, , pp. 223–234.
9. Han, J., & Pei, J. (2000). Mining frequent patterns by pattern growth: Methodology and implications. *SIGKDD Explorations Newsletter, 2*(2), 14–20.
10. Masseglia, F., Cathala, F., & Poncelet, P. (1998). The PSP approach for mining sequential patterns. In *2nd European symposium on principles of data mining and knowledge discovery (PKDD'98)* (LNAI) (Vol. 1510, pp. 176–184). Nantes: Springer.
11. Manilla, H., Toivonene, H., & Verkamo, A. I. (1997). Discovery of frequent episodes in event sequences. *Data Mining and Knowledge Discovery, 1*(3), 259–289.
12. Mooney, C. H., & Roddick, J. F. (2013). Sequential pattern mining: Approaches and algorithms. *ACM Computing Surveys, 45.*
13. Pei, J., Wang, H., Liu, J., Wang, K., Wang, J., & Yu, P. S. (2006). Discovering frequent closed partial orders from strings. *IEEE Transactions on Knowledge and data Engineering, 18*(11), 1467–1481.
14. Pei, J., Han, J., & Mao, R. (2000). CLOSET: An efficient algorithm for mining closed itemsets. In *ACM SIGMOD international workshop on data mining* (pp. 21–30). Dallas: ACM Press.
15. Shyamala, S., & Sathya, T. (2012). Mining closed sequences with constraint based on BIDE algorithm. In *2012 international conference on computer communication and informatics* (pp. 1–5).
16. Vardi, T. (2013, July). *A guide to effective alarm management.* Honeywell Process Solutions. https://www.honeywellprocess.com/library/marketing/whitepapers/Guide-to-Effective-AM-whitepaper_July2013.pdf.
17. Wang, J., Han, J., & Li, C. (2007, August). Frequent closed sequence mining without candidate maintenance. *IEEE Transactions in Knowledge and Data engineering, 19*(8).
18. Zaki, M. J. (1998). Efficient enumeration of frequent sequences. In *7th international conference on information and knowledge management* (pp. 68–75). Bethesda: ACM Press.
19. Zaki, M. J., & Hsiao, C.-J. (2002). CHARM: An efficient algorithm for closed itemset mining. In R. L. Grossman, J. Han, V. Kumar, H. Manilla, & R. Motwani (Eds.), *2nd SSAM international conference on data mining (SDM'02)* (pp. 457–473). Arlington: SSAM.
20. https://www.jointcommission.org/-/media/deprecated-unorganized/imported-assets/jcr/assets/sea_50_alarms_4_5_13_final1pdf.pdf?db=web&hash=5E82688C2EB0B3039476443B834CCF10

Chapter 5
Condition Monitoring of Rotating Machines in Power Generation Plants

Power generation plants are complex systems that have several functional areas such as generators, turbines, bearings, and many more, each of which can in turn be subdivided into components. Turbines are an essential component of power plants and they convert hydraulic power, wind power, steam power, coal power, or nuclear power into electricity. Identifying anomalies at early stages in operating turbines is very important to ensure that these machines do not stop functioning and interrupt power generation to cities. By continuously analyzing sensor data installed in critical components of turbines it is possible to early identify anomalies before they result in a catastrophic equipment stoppage.

Problem Statement

Rotating machines, such as generators, motors, and turbines are very important assets in power generation and industrial applications. Machine reliability and availability are crucial to ensure a reliable power supply. Failure may lead to significant economic losses, due to unexpected outages and possible damage to the asset itself. Monitoring the health of the different components in these rotating machines is of paramount importance to detect potential malfunctions in the equipment and provide proactive/predictive maintenance. The IIoT provides an environment that facilitates remote condition-based monitoring of equipment that then facilitates its proactive/predictive maintenance. Detecting anomalies during the operation of rotating machines is an important approach utilized to monitor the equipment and allows to identify early signs of potential abnormal behavior in one or more components of the piece of equipment. Anomaly detection in a rotating piece of equipment in power generation plants provides a great example of the benefits of the IIoT.

The data generated in real time from hundreds of sensors connected to equipment such as turbines across its functional areas needs to be analyzed in near-real time to detect any anomalies earlier on. Anomalies can originate from various

© Springer Nature Switzerland AG 2021
A. Dagnino, *Data Analytics in the Era of the Industrial Internet of Things*,
https://doi.org/10.1007/978-3-030-63139-0_5

sources and can cause different range of problems. For instance, an anomaly can be overheating of bearing oil and metal components, vibrations from bearings, or low generation of active or reactive power. Hence, it is vital to identify anomaly as soon as it appears. Once anomalies are detected and corroborated, they can be addressed during the scheduled maintenance operations of the equipment.

Background

A hydro-power generation plant has a wide variety of rotating machines such as motors, generators, and turbines among others. These rotating pieces of equipment have important components such as bearing units, rotors, shafts, motor, etc. The rotating machines in power generation plants such as turbines are highly instrumented with a large variety of sensors that collect data at high frequencies on items such as oil temperatures, temperatures of bearings, coil temperatures, metal temperatures, large variety of vibrations and harmonics, active power generated, reactive power generated, among other. These data are generated in real or near real-time and can originate from hundreds of sensors. A set of rotating machines in a power generation plant are interconnected in an IIoT environment and can be continuously monitored to identify any abnormal behavior. The example shown below is based on a real-world rotating machine with data generated at a power plant. The turbine discussed below is a good representative of other turbines in power plants and has the following major components: (a) four bearing systems; (b) a turbine; (c) a generator; (d) a shaft. These components in the rotating machine have over 200 sensors that collect time-stamped data every 10 minutes. The primary time stamped variables collected by the sensors include oil temperatures in the components, different types of metal temperatures in the components (bearings, winding, casings, etc.), pressure values, and a variety of vibrations in the components. Notice that a power generation plant has many rotating machines and the data collected by sensors in these machines is stored in a historian database of an IIoT DCS operating at the power plant. Figure 5.1 shows a block diagram of a typical hydro power generation turbine.

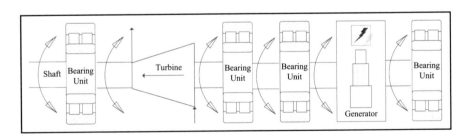

Fig. 5.1 Block diagram of a hydropower generation Turbine

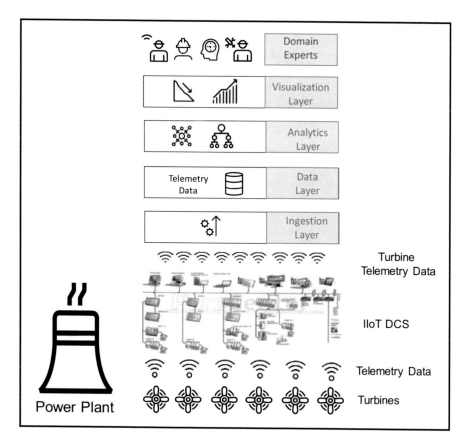

Fig. 5.2 IIoT power generation plant ecosystem

Smart networked products have three primary components: (a) physical/hardware; (b) smart elements, such as sensors, microprocessors, data storage, controls, embedded software and OS, rules engine, gateways, and a UI; (c) connectivity elements, such as ports, antennae, and protocols for wireless and wired communication. Connectivity can be one-to-one, one-to-many-many-to-many. Smart networked products require companies to develop a technology infrastructure that consists of a layered technology stack. This layered technology stack includes the actual smart networked product, embedded software, connectivity capability, Cloud capability to run product software on remote servers, security tools, gateway for external information sources, and a capability to integrate with the enterprise business systems [4].

The turbine system presented above operates in a power generation plant within an IIoT ecosystem of turbines and software and hardware components. Figure 5.2 shows the representation of a power generation plant with networked, instrumented turbines that generate electrical power in an IIoT environment. Figure 5.2 presents different layers like the ones presented in previous chapters that describe an IIoT environment for industrial systems. In this case, a power plant has several systems but an important one is the turbines and generators that use hydro, coal, gas, or nuclear power to generate electricity that electrifies our cities and factories. This is the layer of equipment that is at the bottom of Fig. 5.2. The equipment is highly instrumented with one turbine having up to 200 sensors that capture temperatures of the different components, vibrations and harmonics, power generation output, etc. Depending on the needs, real-time data analyses can be conducted at the turbine level using micro-processors integrated to the turbines. Due to the lower computer capabilities of the micro0-processors, analytic models typically are not as sophisticated as they are in the Cloud and hence, the types of analyses are limited to data filtering or high-level anomaly detection. Nevertheless, the signals and telemetry that originates from the instrumented turbines is then transferred to an IIoT Distributed Control System that stores this data and transmits it to the Data Lake/Data Warehouse through the Data Ingestion Layer.

Moving up in Fig. 5.2 the analytics layer contains a variety of algorithms and for the purposes of this chapter the algorithms that will be discussed conduct anomaly detection of turbines, which allow to identify if a turbine is not working as it is supposed to early on. The visualization layer allows the domain experts to visualize the data being received in the Data Lake/Data Warehouse as well as the anomaly detection results.

Turbine Telemetry Data

An instrumented power generation turbine can have over 200 different sensor readings with a frequency of 5–10 minutes. Hence, the amount of telemetry data collected can be 30,000 to 60,000 data points per day. Typical data attributes that are collected include the ones presented in Table 5.1.

It is important to notice that the units of measurement of attributes have different scales due to the different units in which they are measured. Hence, it is important to normalize these attributes before they can be used by analytic algorithms. The objective of normalization is to adjust the attribute readings to a common scale, typically between 0 and 1. Another important aspect of the data to consider is that many of the data attributes can be highly correlated as in the case of different harmonic readings of a component. Hence, it is necessary to identify these correlated attributes and eliminate them from consideration to simplify the analyses.

Analytics for Anomaly Detection of Rotating Machines

An important advantage of having smart networked machines in industry in an IIoT environment is that they can be monitored utilizing advanced analytic approaches, so that patterns, trends, and early signs of abnormal behavior can be identified

Table 5.1 Data attributes of instrumented Turbines in an IIoT ecosystem

Attribute description	Units
Air pressure	Mbar
Air temperature	Celsius Degrees
Active power	Kilowatts
Current	Amperes
Voltage	Kilovolts
Various temperatures of generator (6 readings)	Celsius Degrees
Speed	g/min
Oil temperature of bearings (various	Celsius degrees
Metal temperatures of bearings (various)	Celsius Degrees
Vertical vibrations for all bearings	Mm/s
Horizontal vibrations for all bearings	Mm/s
Harmonics (various types for al bearings)	Mm/s

before big problems occur and proactive/predictive maintenance can be performed. Analytic models from the Machine Learning discipline can be used to analyze patterns of anomalies in machines and these models can be trained so that can be used on new data for each machine. In addition, it is possible to conduct new types of analytics at the fleet level by identifying patterns across similar machines and increase the number of proactive maintenance and operation activities.

A set of analytic approaches can be integrated to provide valuable insights on potential anomalies in rotating machines such as turbines and in conjunction with domain experts can provide actionable maintenance strategies to avoid costly breakdowns. The analytic methods include the following. *Statistical analysis* provides basic statistics views on the data and help identify frequencies, outlier readings, and missing values. *Unsupervised clustering analysis* helps identify natural clusters in the data vectors being analyzed and the primary data attributes/variables relevant to each cluster. *Anomaly detection analysis* identifies anomalous readings in the values and can point to either faulty sensors, or faulty readings, or real anomalies in the operation of various components in the rotating machine being analyzed. *Feature selection and prediction analysis* allows to identify the most salient variables in a highly multi-dimensional analytics space and develop a predictive model for anomalies. The combination of all four types of analyses provides a robust way to analyze the condition of rotating machines such as turbines of the type described here. Figure 5.3 shows the analytics models sequence required to analyze the data of a rotating machine such as a turbine and include: (a) statistical analyses; (b) unsupervised clustering analysis; (c) anomaly detection analyses; (d) prediction analysis.

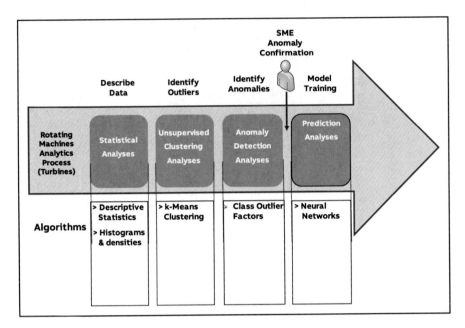

Fig. 5.3 Analytic process to identify anomalies in a rotating machine such as a Turbine

Statistical Analysis of Turbine Data

Statistical analysis helps the domain expert to identify for each data attribute a histogram of values, minimum values, maximum values, standard deviations, average values, missing or null values, etc. This type of analysis is very useful to domain experts to quickly visualize if sensor values are within expected ranges and to identify possible extreme outlier values. Another important aspect of conducting basic statistical analysis is to identify areas where there is missing data due to failure in sensor readings or data recording. Statistical analysis helps the domain expert to cleanse the data and prepare it for further analyses. Figure 5.4 shows a diagram that displays a portion of the results of statistical analysis conducted on the turbine sensor data.

Clustering Analysis of Turbine Data

The k-means clustering ML algorithm determines a set of k clusters and assigns each data observation to exact one cluster. The clusters consist of similar observations and the similarity between observations is based on a distance measure between them [11]. The objective of this analysis is to identify clusters of significance in the turbine dataset, especially to identify any outlier values in the dataset.

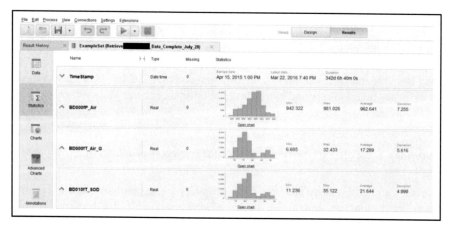

Fig. 5.4 Sample statistical analysis of Turbine Data

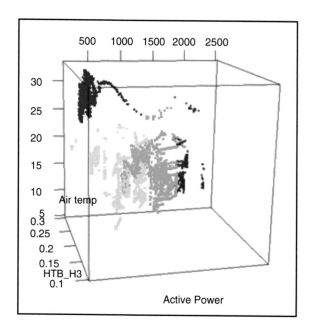

Fig. 5.5 Clustering analysis of Turbine data using k-Means clustering ML algorithm

These clusters share certain common characteristics on the time stamped turbine readings. Figure 5.5 shows four main clusters that form in the analysis of the dataset. The blue cluster consists of few observations that seem to be outlier values and could possibly be incorrect measurement/observations in the system. The yellow, green and red clusters are observations that for the most part fall within the expected values, although there may be some observations that could be anomalies in the data, but with this analysis these anomalies cannot be corroborated.

It is important to point out that as usual, the subject matter expert, or Service Engineer, needs to review the results of the clustering analysis to verify and validate that clusters are either within the range of accepted values or are outliers. For the purpose of this analysis outlier values are those that are observations that are far away from the mean or median in a distribution. Anomaly values are values that are not as far away as outliers and can indicate abnormal behavior in a system.

Anomaly Detection Using Connectivity-Based Outlier Factor

Typically, and unless there is a very serious issue, power generation equipment in a power generation plant operates without showing any visible abnormality. For this reason, it is important to have robust analytic approaches to identify any abnormality in the components of rotating pieces of equipment as soon as possible to conduct proactive maintenance and avoid costly stoppages of the equipment (loss of power generated, legal fees, high corrective maintenance costs, high labor costs for unplanned maintenance, etc.). Abnormal behavior in a smart networked rotating machine such as the one described can originate from the overheating of the oil inside any of the bearing systems, or overheating of the copper in the generator, or excessive vibrations in a shaft or bearing unit. Identifying abnormal behavior from these systems is difficult due to the large amount of data generated from frequent readings (high frequency) and a multitude of sensors where this data originates (high dimensionality on the data).

One way to address analyzing high frequency and high dimensional sensor data to identify abnormal behavior of rotating equipment, is by using "Anomaly Detection" methods, which belong to "Unsupervised Machine Learning". Unsupervised analytic methods do not have a training dataset with observations labeled as "abnormal". Anomaly detection algorithms originate from "outlier detection" models [15]. Outlier detection analytic models can be grouped into four major schools of thought: (a) distance and density based methods [8]; (b) subspace based methods [15]; angle based methods [6]; and (d) ensemble based methods [15].

There are many algorithms that can be used for anomaly detection such as the Connectivity-based Outlier Factor (COF) by Tang et al. [13] and the Local Outlier Factor by Breunig et al. [2]. Abnormal values can be outliers (values that are markedly numerically distant from the rest of the observations and in most cases point to a direct measurement error) or anomalies (values that are distant from the rest of the observations but are feasible). Domain experts are very well equipped to distinguish between outlier values and anomalies and for this reason, it is important that these domain experts make the final determination on anomalies, so they can be labeled as such. Figure 5.6 shows sample abnormal vector entries in the turbine data analyzed using the COF algorithm. The entries shown in the "Outlier" column that are greater than 1.0 are considered anomalies and these are analyzed by the domain expert to confirm them as anomalies.

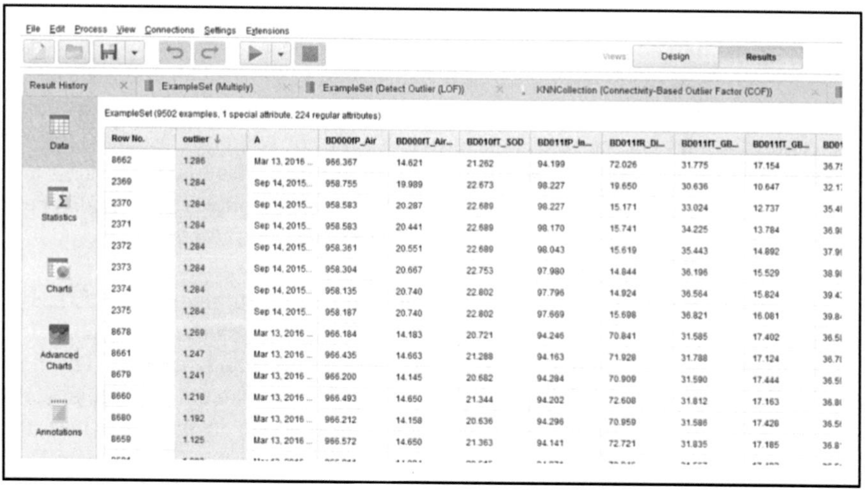

File Edit Process View Connections Settings Extensions

Views   Design   Results

Result History   × ▦ ExampleSet (Multiply)   × ▦ ExampleSet (Detect Outlier (LOF))   ×   ▪ KNNCollection (Connectivity-Based Outlier Factor (COF))   ×

ExampleSet (9502 examples, 1 special attribute, 224 regular attributes)

Row No.	outlier	A	BD000IP_Air	BD000IT_Air...	BD010IT_SOD	BD011IP_In...	BD011IR_DI...	BD011IT_GB...	BD011IT_GB...	BD0
8662	1.286	Mar 13, 2016	966.367	14.621	21.262	94.199	72.026	31.775	17.154	36.7
2369	1.284	Sep 14, 2015	958.755	19.989	22.673	98.227	19.650	30.636	10.647	32.1
2370	1.284	Sep 14, 2015	958.583	20.287	22.689	98.227	15.171	33.024	12.737	35.4
2371	1.284	Sep 14, 2015	958.583	20.441	22.689	98.170	15.741	34.225	13.784	36.9
2372	1.284	Sep 14, 2015	958.361	20.551	22.689	98.043	15.619	35.443	14.892	37.9
2373	1.284	Sep 14, 2015	958.304	20.667	22.753	97.980	14.844	36.196	15.529	38.9
2374	1.284	Sep 14, 2015	958.135	20.740	22.802	97.796	14.924	36.564	15.824	39.4
2375	1.284	Sep 14, 2015	958.187	20.740	22.802	97.669	15.696	36.821	16.081	39.8
8678	1.269	Mar 13, 2016	966.184	14.183	20.721	94.246	70.841	31.585	17.402	36.5
8661	1.247	Mar 13, 2016	966.435	14.663	21.288	94.163	71.928	31.788	17.124	36.7
8679	1.241	Mar 13, 2016	966.200	14.145	20.682	94.294	70.909	31.590	17.444	36.5
8660	1.218	Mar 13, 2016	966.493	14.650	21.344	94.202	72.608	31.812	17.163	36.8
8680	1.192	Mar 13, 2016	966.212	14.158	20.636	94.296	70.959	31.586	17.426	36.5
8659	1.125	Mar 13, 2016	966.572	14.650	21.363	94.141	72.721	31.835	17.185	36.8

Data

Σ Statistics

Charts

Advanced Charts

Annotations

Fig. 5.6 Anomaly detection for Turbine data using COF ML Algorithm

Once the domain expert confirms the abnormal vectors, it is then possible to compare the values of new observations of the turbine and determine whether the new values correspond to normal operation of the turbine or if there is an anomaly in the operation of the turbine. A Neural Network could be trained with the results of the anomaly detection outcomes and then new values from a turbine can be filtered through the Neural Network to define if the turbine is operating normally or not.

Enhancing Domain Knowledge of Power Engineers Through Anomaly Detection System

An IIoT OEM organization that manufactures equipment such as power generation turbines, control systems, and services to customers such as power generation plants can greatly benefit from providing remote service offerings to its customers. Remote services, built in an IIoT platform, provide an opportunity for a new revenue stream and changes the business model to provide proactive/predictive services instead than becoming reactive and diagnostic. Service Engineers that diagnose and repair power generation turbines are a very valuable resources, in high demand, and very scarce. By using a remote anomaly detection IIoT system for turbines that continuously scans sensor data and learns from this historical sensor data, knowledge about operating anomalies of turbines is captured and then re-used through analytic models. This approach provides high-knowledge and consistent anomaly detection service capabilities to the OEM service provider, which keeps customers ahead of potential catastrophic failures of their equipment. Remote anomaly detection capabilities help to hire less experienced service Engineers that can perform at par with their more experienced counterparts by using the advanced analytics system.

The analytic approach described to identify anomalies in the operation of power generation turbines can be re-used in different types of equipment such as hydro-power generation turbines, gas turbines, and coal turbines. The analytic process and models learn from the historical data collected for each turbine. This can also help to cross-pollinate the knowledge among service Engineers and increase the pool of experts.

Remote monitoring and analytics of historical data of turbines using a remote monitoring system as described above is an important first step towards changing the traditional business model which is more diagnostic and reactive. The new approach is more proactive and predictive and relies less on an on-site specialist. In the traditional model, Service Engineers are typically sent to the site where the turbine(s) operate to extract the data and perform the analyses. In this new IIoT approach, Service Engineers can work from a central location and hence the travel needed is minimized. As explained, this step brings the turbine manufacturer/provider with the capability that in the future instead of selling the equipment and service it, the manufacturer will be able to sell power generated per unit of time, and the utility will not need to be concerned to service the equipment. Similarly, the IIoT OEM could allow the utility to only re-configure electronically the requirements of the turbine and the equipment will change its behavior based on the utility's request. This new approach helps generating new types of jobs that are more knowledge-intensive and focused on predictive analytics. These new jobs are also more satisfying to service Engineers as they can help taking the burden of prolonged traveling away.

References

1. Abdulla, K., & Harous, S. (2016). Internet of things: Applications and challenges. In *12th IEEE International Conference on Innovations in Information Technology (IIT)* (pp. 212–217).
2. Breunig, M. M., Kriegel, H. P., Ng, R. T., & Sander, J. (2000). LOF: Identifying density-based local outliers. In *Proceedings of the ACM SIGMOD 2000 International Conference On Management of Data* (pp. 1–12) Dallas.
3. Guyon, & Elisseeff, A. (2003). An introduction to variable and feature selection. *Journal of Machine Learning Research, 3*, 1157–1182.
4. Porter, M. E., & Heppelmann, J. E. (2015, October). How smart connected products are transforming companies. *Harvard Business Review, 93*, 4–19.
5. Porter, M. E., & Heppelmann, J. E. (2014, November). How smart connected products are transforming competition. *Harvard Business Review, 92*(11), 4–23.
6. Gubbi, J., Buyya, R., Marusic, S., & Palamiswami, M. (2013, September). Internet of things (IoT): A vision, architectural elements, and future directions. *Future Generation Computer Systems, Elsevier, 29*(7), 1645–1660.
7. Knox, E. M. & Ng, R. T. (1998). Algorithms for mining distance based outliers in large data-sets. In *Proceedings of the 24th International Conference on Very Large Databases* (pp. 392–403). Citeseer.

8. Kriegel, H. P., Schubert, M., & Zimek, A. (2008). Angle-based outlier detection in high-dimensionality data. In *Proceedings of the 14th ACM SIGKDD International Conference on Knowledge Discovering Data Mining* (pp. 444–452). ACM.
9. Perera, C., Liu, C. H., & Jayawardena, S. (2014, October). The emerging internet of things marketplace from an industrial perspective: A survey. *IEEE Transactions on Emerging Topics in Computing, 3*(4), 585–598.
10. Reinfurt, L., Breitenbucher, U., Fankenthal, M., Leymann, F. & Riegg, A. (2016). Internet of things patterns. Europlop '16: *Proceedings of the 21st European Conference on Pattern Languages of Programs*, ACM.
11. Salman, R., V, K., Li, Q., Strack, R., & Test, E. (2011, July). Fast k-means algorithm clustering. *International Journal of Computer networks and Communications, 3*(4).
12. Satyanarayanan, M., Simoens, P., Xiao, Y., Pillai, P., Chen, Z., Ha, K., Hu, W., & Amos, B. (2015). Edge analytics in the internet of things. *IEEE Pervasive Computing, 14*(2), 24–31.
13. Tang, J., Chen, Z., Fu, A. W., & Cheung, D. (2002). Enhancing effectiveness of outlier detections for low density patterns. In *Pacific-Asia Conference on Knowledge Discovery and Data Mining* (pp. 535–548).
14. Xu, L. D., He, W., & Li, S. (2014, November). Internet of things in industries: A survey. *IEEE Transactions on Industrial Informatics, 10*(4), 2233–2243.
15. Zhang, K., Hutter, M., & Jin, H. (2009). A new local distance based outlier detection approach for scattered real-world data. In *Advances in knowledge discovery and data mining: Proceedings of the 13th Pacific Asia conference on knowledge discovery and data mining* (pp. 813–822). Springer.
16. Zimek, E. S., & Kriegel, H. P. (2012). A survey on unsupervised outlier detection in high-dimensional numerical data. *Statistical Analysis and Data Mining, 5*(5), 363–387.

Chapter 6
Machine Learning Recommender for New Products and Services

A global OEM enterprise wishes to use its IIoT infrastructure to identify opportunities of new sales, up-sales, and cross-sales of their products and services within their customer installed base. The IIoT enterprise wishes to increase the share of wallet of their customer base. From the perspective of data analytics, it is possible to identify similar customers that buy baskets of products from the OEM and identify gaps on the baskets that can be addressed by offering new products and services. It is feasible to achieve this analytic capability using ML techniques such as Classification Analysis, Market Basket Analysis and Association Rule Mining, and Sentiment Analysis.

Problem Statement

The goals of approach include developing and implementing a system capable of analyzing data originating from the sales of products and services and execution of services. These data are typically stored across various systems such as Customer Relationship Management, Customer Complaint Resolution, Enterprise Resource Planning, and Service Support systems to efficiently monitor various KPIs related with product sales, service sales, service execution, customer purchase, customer purchase preferences, installed base management, customer demographics, customer satisfaction. The analytic approaches that are implemented are expected to derive KPI's that will be predictive in nature and will be derived using data mining and machine learning algorithms. Ultimately, the objective is to have a Machine Learning-based system that provide recommendations to Sales Domain Experts and Service Sales Engineers on new products and services that can be sold to their customer base that are not currently buying them. The ML functionality aims at analyzing the existing set of customers who are regularly engaging with IIoT organization and buying products and services (the company's customer base), to identify what might be other products or services to be sold to these customers.

© Springer Nature Switzerland AG 2021
A. Dagnino, *Data Analytics in the Era of the Industrial Internet of Things*,
https://doi.org/10.1007/978-3-030-63139-0_6

Background

As part of its digital roadmap an IIoT OEM organization began a journey to utilize advanced analytics and machine learning techniques to help increase service revenues of its manufacturing and service businesses. As part of this digital initiative, the plan has been to build a robust and powerful AI framework that can utilize historical data being generated from product and service sales to provide actionable insights into customer requirements thereby unearthing hidden opportunities for sale of services and thereby increasing customer satisfaction, conversion ratio and the overall service business. From the existing set of the IIoT organization's customers the objective is to identify potential to sell service contracts and/or additional products and services based on their current purchases and satisfaction score with the OEM organization.

For operational purposes, the IIoT organization uses several data repositories that are relevant to this ML application and are shown in Fig. 6.1. These data repositories are shown at the bottom of Fig. 6.1 and grouped in the Service, Product and Customer data layer. The first data repository is the **Sales Opportunities** database that maintains information on all bids sent to potential customers in all their stages,

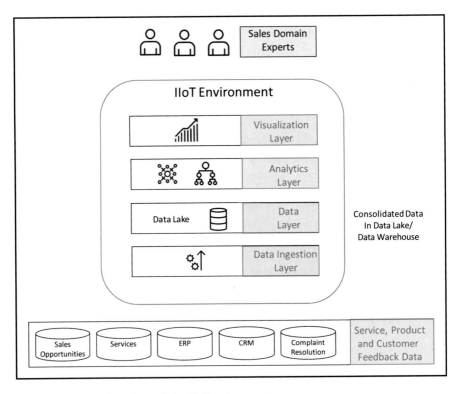

Fig. 6.1 Product and service analytics IIoT environment

tracks won and lost bids, and tracks all sales to the company's customers. This data repository has information feedback when a bid has been won by the company and if the bid was lost the potential reasons why the bid was lost. Another data repository relevant to this application includes the **Services and Products** database that contains information on the types of services provided to the customers of the IIoT organization. Services are typically provided to customers that buy products but, in some occasions, services can be sold to customers that buys competitor's products and get them services by the IIoT organization. **Enterprise Resource Planning** data repository contains information about products that have been purchased by the IIoT customer base and all the purchase and production history. The **Customer Relationship Management** data repository contains regular feedback from the IIoT customer base on their level of satisfaction with the interactions with the company, the level of satisfaction on the products they buy from the company, and the level of satisfaction on the services they purchase. Feedback is typically provided at the end of a transaction with the customer and regularly once or twice a year. This database contains a combination of categorical data as well as free text feedback provided by the customers. Finally, the customer **Complaint Resolution** database contains feedback from customers that have raised any complaint on the products or services provided by the organization. Complaints include late deliveries, missing parts, defective products, failures in products during warranty period, and other complaints the customers may raise during their interaction with the IIoT company. This database contains a mixture of categorical and textual information that is generated by the customer and the IIoT organization during the lifecycle of the customer complaint, from entering the complaint to the final resolution.

Moving up in Fig. 6.1 we observe that the IIoT organization's environment is also shown as a layered diagram. The Data Ingestion Layer is a set of applications that extract the data from the operational data repositories using the data model required to run the Service and Product recommender application. The raw data is ingested and stored in the "staging area" of the Data Lake or Data Warehouse. Once in the landing area, the data can be pre-processes and made ready to be analyzed to produce the expected recommendations. The ML analytic models used to produce these recommendations include the **K-Means Clustering** ML algorithm to classify similar customers into groups. Another algorithm contained in this layer is the **Association Rules Mining** ML algorithm that performs market basket analysis to identify what customers in each cluster typically buy to identify gaps or new opportunities among customers in each cluster [3].

Another ML algorithm used is **Text Mining Sentiment Analysis** to evaluate how a customer feels about the IIoT organization. Sentiment analysis is conducted in three different areas. First, sentiment regarding the interaction that the customer has had after a purchase of a product or service. Second, sentiment regarding the overall relationship the customer has with the IIoT organization. Third, sentiment regarding the way the customer feels on how the IIoT organization addresses customer complaints or customers concerns. Another algorithm uses the KPIs generated by the clustering, market basket analysis, and sentiment analyses algorithms to produce an overall win-loss recommendation to the Sales Domain Experts.

Table 6.1 Data attributes and units of historical sales of a product

Operational data repository		Data attribute	Data type
Sales database		Customer ID	Categorical
		Customer location	Categorical
		Product enquired	Categorical
		Won/loss data and dates	Categorical & Date
		Deal Value	Numeric
Services		Customer/site ID	Categorical
		History of services acquired and dates	Categorical & Date
		Site location	Categorical
Enterprise Management	Resource	Customer ID	Categorical
		History of products purchased and dates	Categorical
Customer Management	Relationship	Customer feedback	Textual
		Customer rating	Categorical
Complaint Resolution		Customer feedback on complaints	Textual
		Customer rating on complaints	Categorical

The visualization layer is also quite important to translate the different KPIs into information that is readily absorbed by the Sales domain experts. Through the visualization layer, the Sales domain experts analyze the recommendations the system provides and based on their knowledge on the IIoT organization and their customers, make the appropriate decisions to qualify sales opportunities. Once these sales opportunities are appropriately qualified, the Sales domain experts approach customers in their portfolio to offer new products or services that are relevant to the customers.

Historical Data

To conduct the different analyses discussed in the section above, several data attributes are required. These data attributes are extracted from the organization's operations databases (Sales Opportunities, Services, ERP, CRM, and customer Complaint Resolution). Table 6.1 shows the database source, the data attributes used, and data types that are needed for the analytics application.

Product and Services Recommender Analytics

Figure 6.2 shows the analytics process required to produce the recommended product or services to customers. Moving from left to right in the diagram, the original operational data repositories used to extract the data required for the application are

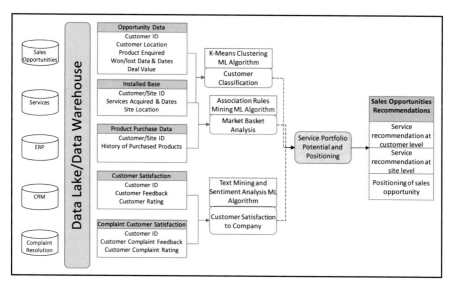

Fig. 6.2 Service and product recommender analytics process

stored in the IIoT application Data Lake/Data Warehouse. The data attributes used to conduct the analytics include Opportunity Data, the Installed Base data, Product Purchased Data, Customer Satisfaction data, and Complaint Customer Satisfaction data. With the Opportunity Data, the Installed Base data the Customer Classification Analytics is conducted using the K-Means Clustering ML algorithm. With the Installed Based data and Product Purchase Data the Market Basket Analysis is conducted using the Association Rules Mining ML Algorithm. Finally, with the Customer Satisfaction data, and Complaint Customer Satisfaction data the sentiment of the customer towards the company is determined by using the Text Mining – Sentiment Analysis ML algorithm. With these three analyses, then a set of ranked sales opportunities are recommended by the system to the Sales domain experts to propose to the customers in their portfolio.

Customer Classification Analytics

Customers in the installed base of the IIoT company are classified in clusters based on the purchases they have made over time or based on purchase patterns. The K-Means Clustering ML algorithm is used, and the clusters are created using the Product Type. The clusters or groups are not known a priori because an unsupervised clustering approach is utilized, and a number of "k" clusters is formed based on the historical data.

		Product 1	Product 2	Product 3	Product 4	Product 5
	Customer A	X				X
	Customer B		X		X	
	Customer C					X
	Customer D		X	X		
	Customer E			X		
		Product 1	Product 2	Product 3	Product 4	Product 5
Cluster I	Customers B, D, E		X	X	X	
Cluster II	Customers A and C	X				X

Fig. 6.3 Sample of customer clustering based on purchasing patterns

Figure 6.3 shows a simple example on how this clustering occurs. A set of five customers for the installed base and they have bought a variety of products available that go from Product 1 to Product 5. Using the data in the top matrix, two clusters are created. Cluster I contains customers that have bought Products 2, 3, and 4. Cluster II contains customers that have acquired Products 1 and 5.

Market Basket Analysis

The Association Rule Mining ML approach is used to identify gaps in products and services that customers in a certain cluster can have. Association rule mining is a machine learning model that discovers interesting relationships among variables that contain large volume of historical transactional data [1, 4]. A classic use of Association Rule Mining is to analyze transactional data in a point-of-sale process in supermarkets. The idea is to analyze the patterns of purchases of customers and generate rules of frequent buying observations. Association Rule Mining assumes all data is categorical and was initially used for market basket analysis to find how items purchased by customers are related. Market basket analysis derive rules like a customer that buys ham, sliced cheese, and tomatoes, is likely to buy bread {ham, sliced cheese, tomatoes} → {bread}.

Figure 6.4 shows how the Association Rule Mining ML algorithm is used to identify potential gaps in the customers that belong to a cluster. These gaps are the ones that show potential sales opportunities once they have been properly positioned.

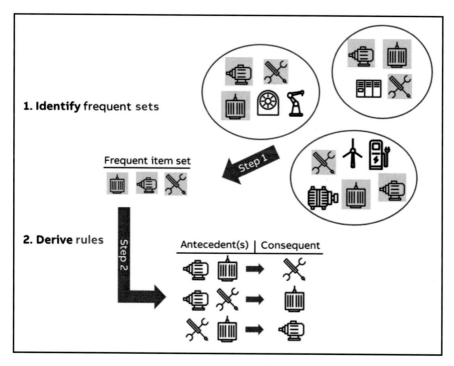

Fig. 6.4 Sample market basket analysis

Sentiment Analysis

Feedback from customers about interactions that have had with the IIoT organization are provided in text form and stored in the CRM data repository. Similarly, feedback from customers on how the IIoT organization has addressed any potential complaint are also provided in text form and stored in the Customer Complaint database. These text responses from customers can be analyzed using Sentiment Analysis that is the practice of applying Natural language Processing and Text Analysis techniques to identify and extract subjective information from a piece of text [2]. Sentiment Analysis is the process of analyzing unstructured text to extract relevant information and determine if an expression is positive, neutral, or negative and to wat degree. This approach analyzes and measures human emotions and convert them into "hard facts". Sentiment Analysis is a text analytics approach that provides insight into the emotion behind words and can provide an organization with specific feedback about how they feel about the organization and its products and services [5].

The Sentiment Analysis algorithm processes textual data and performs sentence splitting and tags the sentences based on polarity and intensity of sentiments. With this information the algorithm provides its output reporting the "sentiment" the text. The Sentiment Analysis ML algorithm has three primary stages. First, the algorithm processes the textual data and performs sentence splitting. At this stage, non-textual data is eliminated. Each sentence is examined for subjectivity. Sentences with subjective expressions are retained and the ones with objective expressions are discarded.

Second, the algorithm tags each split sentence based on their polarity and the intensity of the sentiments in the sentence. Sentiments can be broadly classified into two groups, positive and negative. Each subjective sentence is classified into positive, negative, good, bad, like, dislike. The algorithm scans keywords to categorize a sentence as negative or positive based on a simple binary analysis, i. e. "enjoyed" = good; "miserable" = bad.

Finally, the algorithm provides the results of the analysis. The results are converted into values associated with the classification of the sentences. Algorithms classify sentiments into negative, neutral, or positive and a score can be associated so is possible to use for further computation. For example, a customer statement given in the CRM system such as: "We were very happy that the spare part required to fix our conveyor belt was delivered fast" is considered as positive feedback and given a score of 3.0. A feedback such as "The repair of our conveyor way completed" can be considered as neutral and a score 2.0 is given. While a sentence: "we were not happy with the delivery time of the spare part of our conveyor because we lost half day productivity" is viewed as negative and hence given a score of 1.0. The final sentiment of the customer then is calculated by averaging all inputs and deriving either a positive, neutral or negative view of the IIoT organization. Sentiment is calculated for the following items: (a) the overall sentiment of a product sold to all customers for all products; (b) the overall sentiment of a service sold to all customers, for all services; the sentiment of the customer towards the IIoT organization in general terms of engagement; (d) the sentiment of the customer towards how the IIoT organization has addressed customer complaints in the past.

With the values of the different sentiments, when the sales domain expert views a recommendation to offer a customer a service such as "Calibration of Rotor in Electric Motor", the domain expert will identify if the customer is a promoter of the IIoT organization (positive sentiments) or a detractor (negative sentiments). The Domain expert will also view if the general sentiment of all customers that have been offered the calibration service view it as positive or negative. And based on her/his knowledge on the customer will be able to decide if this potential sales opportunity has potential.

Enhancing Domain Knowledge of Service Engineer Salespeople Through the Product and Services Recommender System

An IIoT analytic system, similar to the one described above, that provides recommendations of new potential sales, up-sales and cross-sales of products and/or services facilitates the exchange of business expertise among salespeople and enhance their development of sales opportunities in their customer portfolio by increasing pollination of their domain knowledge. Figure 6.5 below shows a "sanitized and anonymized" dashboard of the recommender system described above that contains the KPIs and information the product and service salespeople can utilize to guide them to make specific recommendations to customers in their portfolio. The dashboard contains the results of the analytics described in this chapter. Starting from the top of the dashboard, there are windowpanes that present the "sanitized name' and information of the customer that a specific service salesperson has analyzed to plan her/his recommendations. Moving down to the middle panes of the dashboard shown in Fig. 6.5, the middle and left-hand side pane shows the results of the sentiment analyses conducted. First, the "sanitized customer ID" is shown, and then three sentiment analyses scores are shown, where scores 2 or greater than 2 are considered positive. The customer satisfaction score (NPS), or sentiment of the customer' satisfaction in their interactions with the IIoT OEM is shown to be 2.66, which means quite positive. Moving towards the right, the score of customer's

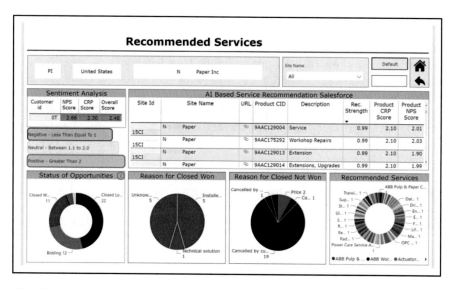

Fig. 6.5 Sample dashboard view for marketing recommender system

sentiment on the quality of products or services that the OEM has globally provided shows to be quite positive as 2.33. The overall sentiment of the customer towards the OEM is the average and shows a very positive score of 2.48. On the far-right pane in the middle of the dashboard, the list of services recommended by the system are shown to the service salesperson. The sentiment scores of other customers that purchased the recommended services towards the OEM in terms of quality and overall responsiveness/interaction are shown in the two columns on the right of the pane. The strength of the recommendation in terms of confidence level that the customer will receive these recommendations positively is also shown. Finally, in the lower part of the dashboard, general information of the customers history in terms of other purchases made is shown to the service salesperson. A status of opportunities and sales is shown as well as a pie chart that summarizes the recommended services that the customer may be open to buy.

The analytic system described above has brought high value to the salespersons in the OEM. The system allows salespeople and service Engineers to proactively create sales opportunities for the customers in their portfolios, that are more qualified as these opportunities are generated based on knowledge on customers' buying patterns and their sentiment toward the OEM. The system facilitates the OEM to provide its customers with opportunities to improve their operations and potentially to save money on costly repairs and stoppages in their operations due to equipment failures or to acquire more efficient products for their production. The sales opportunities identified by the IIoT system are considered of "high quality" and allow salespeople and service Engineers to use their deep domain expertise to further qualify these opportunities and increase the hit rate probability. This provides both the salespeople and the customers with high level of satisfaction.

Having a system like the one described provides an opportunity to increase consistency on recommendations provided to globally distributed salespeople and increases the level of cross-pollination on the knowledge they possess. The system does not make decisions for the salespeople but empowers them to make more robust and data-driven decisions on their offerings to their customers. As sales opportunities increase, an unexpected benefit is the hiring of new salespeople into the organization to manage the new opportunities created. Less experienced or new salespeople can also utilize the system to generate and explore recommendations for new sales, cross-sales, and up-sales to customers with high degree of confidence, which will provide the opportunity to interact with their more experienced counter partners and learn from them at the same time of servicing the customer. Hence, new and less experienced product and service salespeople can achieve high level of performance and consistency like their more experienced counterparts.

References

1. Agrawal, R., Imielinski, T., & Swami, A., (1993). Mining association rules between sets of items in large databases. In *Proceedings of the 1993 ACM SIGMOD International Conference on Management of data* (pp. 207–216).
2. Kauffmann, E., Peral, J., Gil, D., Ferrandez, A., Sellers, R., & Mora, H. (2019). Managing marketing decision-making with sentiment analysis: An evaluation of the Main product features using text data mining. *Sustainability Journal, 11*(15), 1–19.
3. Srinivas, B., Ramesh, G., & Sriramoju, S. B. (2018). An overview of classification rule and association rule mining. *International Journal of Scientific Research in Computer Science, Engineering and Information Technology, 3*(1), 1692–1697.
4. Surjandari, I., & Seruni, A. (2005). Design of Product Placement Layout in retail shop using market basket analysis. *Makara Journal of Technology, 9*(2), 43–47.
5. https://www.slideshare.net/AngieTabone/sentiment-analysis-78199104

Chapter 7
Managing Analytic Projects in the IIoT Enterprise

The development of analytic projects in IIoT environments requires a robust framework, sound Software Engineering processes and methods, and a flexible development lifecycle methodology. This chapter discusses all three: a development framework, key Software Engineering processes, and a flexible development methodology lifecycle required to develop analytic applications that can be used to unleash the power of IIoT platforms.

Managing data analytics projects, such as the ones required to propel the Industrial Internet of Things requires skilled project management practices. Analytic projects have all the same characteristics of software development projects, but they have some special elements that require special attention. First, it is necessary to develop software capabilities to extract, ingest, cleanse, and pre-process the historical data that will be used to derive the required diagnostic, predictive, and prescriptive metrics to help decision making of SMEs. Second, the functionality expected from analytic projects is directly related with the algorithms used (statistical models, data mining models, or machine learning models) and the concatenation of these. It is then imperative to clearly describe the requirements of the system so that the results from the analytic models can be properly verified and validated. This is important because several of these models, especially the ones that provide a diagnostic, or a prediction, or a forecast, or a prescription, typically learn from historical data and provide results that are mathematically correct but in terms of the real world they may have a degree of deviation. This leads us into having to define a robust **Requirements Engineering** process to be able to define in a precise manner the requirements that are to be achieved in an analytics project. Having a good definition of the analytic requirements is essential to identify precise ways on how to **Verify** and **Validate** the results of predictions, forecasts and/or prescriptions of analytic systems. Moreover, the **Project Management** process will help manage the development lifecycle that will ensure the delivery of the system as expected by the customers/users [4].

Another aspect that is important in analytic solutions is the way in which the results are presented to the user and how the user interacts with these results.

© Springer Nature Switzerland AG 2021
A. Dagnino, *Data Analytics in the Era of the Industrial Internet of Things*,
https://doi.org/10.1007/978-3-030-63139-0_7

This part of the analytics functionality is achieved via the visualizations and dashboards developed for the system. Analytic packages such as R-Studio or Python, Azure ML, or analytic shells such as Rapid Miner or Weka, have libraries that visualize the results of analytic models. But they often need to be either re-developed or imported in some fashion into visualizations modules of IIoT systems to tailor them to the domain experts or SMEs and provide the interaction capability with the expected performance, usability, availability, security, and other quality attributes.

The Project Management process for analytic applications is crucial to ensure that the project delivers the expected results "quickly", with the expected functionality, on time and on budget. It is typically expected that the development team closely interacts with the system users and sponsors to show results and progress fast and be able to receive continuous feedback from users and incorporate it into the development. For this reason, the use of agile development lifecycles such as Scrum [13] or Agile Kanban [2] are preferred in the development of advanced analytic solutions for the IIoT enterprise.

Definition Phases of an Analytics Project in the IIoT Enterprise

The results of analytic projects need to be clearly understood by the receivers and users of the system. The analytics solutions need to demonstrate value in the organization and truly affect change so that the system get the appropriate recognition in the organization. Successful analytics in the IIoT organization requires of course deep knowledge on the analytics domain to understand how models can be used to address different business situations. Moreover, it is also one way to achieve this is by aligning the steps of the data analysis lifecycle to the traditional steps of business management [14]. The general phases required to define an advanced analytics project include: (1) development of a business case; (2) analysis of business benefits; (3) relate results to benefits; (4) actions required to realize benefits; (5) implementation; (6) completion. Analytics domain experts need to both be competent in the data science aspect and be able to incorporate their analytics expertise into the business management cycle. The data analysis domain expert needs to bridge the data analytics lifecycle to the six different phases of the business management lifecycle.

Define the Business Case A successful analytics project needs to provide the organization with real and actionable insights. In enterprise analytics, this is known as the business question. This refers to identifying the reason behind the project, what business problem is going to be addressed. By thinking about a business question in terms of a business case, pain point to resolve, or opportunity to address it will be possible to better explain to key stakeholders exactly why an analytics project is important.

Preliminary Benefit Realization Analysis Once the estimated benefit has been defined, the next step is to try to quantify the benefit by analyzing the key performance indicators that the analytics project will address. This analysis is aimed at explaining what benefit(s) the IIoT organization will derive from the project in a more quantitative manner. The estimated benefits include increased revenues, increase in customer retaining, reduction on operating costs, increase in market penetration and hence increased revenues, increased customer satisfaction, etc.

Interpretation As data are collected for the analytics project and begins to be profiled and analyzed, it is important to keep in mind the alignment with the original business goals. This is to ensure that the analytics activities are delivering on the expected business benefits. During this phase of the project, a realignment on the objectives and business benefits may take place. For example, new business benefits may be identified and hence they should be added to the original business goal(s). It is also important during this phase, to begin thinking on specific actions that can be derived from the analyses to achieve the business objectives. Phrasing the interpretation in terms such as, "Looking at the data, we can see X and Y, which may explain [the business question]" can be very effective in allowing the analyses to translate into actionable steps for the IIoT organization.

Definition of the Project Scope In this phase it is important to clearly define the analytics project and its scope to create a work breakdown structure with the tasks related to the analytics project. That is why is just as important to define what we are not going to do as it is to define what we are going to do in the project so that both the team and stakeholders have a clear understanding before the project gets too far along. By thinking about realization in the terms of scope, it will be easier for the team to tie its interpretation to next steps and offer guidance to the organization moving forward. This is one of the best ways that the team can affect change in how the IIoT organization will do business.

Once the analytics project has been defined, then, the project is implemented. This is the phase when the analytics project is carried out. From the analytics perspective it is essential to keep in mind the KPIs that are to be determined and are linked to the business goals of the project. From the project management, this is the phase when the Project Manager monitors controls the project to ensure on time delivery. While the analytics team should focus on the KPIs and data points that are most relevant to the data project, it may be beneficial to again think back to the business question when communicating this information to key stakeholders. If they don't understand why a data point is important, help them understand by explaining it in business language. Once the project has been implemented, it is important to link it to the business benefits defined earlier in the project scoping. A retrospective analysis needs to be conducted to identify lessons learned, actions the IIoT organization should take to move forward, what went right and what went wrong, and what should be done differently next time. "By following suit, you can hold some real influence over the strategic next steps that an organization takes, in terms of both future analytics projects as well as business initiatives" [1].

Delivery Framework for IIoT Advanced Analytics Projects

The development of analytic applications for the IIoT enterprise requires a robust, stable, and reliable development framework to ensure consistency, quality, and on-time and on-budget delivery that delights the expectation of the IIoT enterprise users. Figure 7.1 shows a high-level development framework that can be used by IIoT enterprises to develop successful analytics applications.

The development framework has three phases as shown in Fig. 7.1. First, sustaining activities are foundational activities that are required to define the analytics project in an IIoT organization. From the software development process, there are four primary process areas that play a pivotal role in the foundational activities and include Requirements Development and Requirements Management (Requirements Engineering), Project Planning and Project Monitoring and Control (Project Management). These process areas need to be robustly implemented in the organization to develop sustainable IIoT analytic solutions.

Sustaining Phase

The sustaining phase provides the foundational elements of any advanced analytics project for the IIoT enterprise. This phase is subdivided into four distinct stages: (a) Demand Management; (b) Scope; (c) Requirements and Design; (d) Project Planning. This phase is shown in Fig. 7.2.

An IIoT enterprise defines the business need for an analytics project during the *Demand Management stage*. The business need is derived from the identification of a pain point in the organization, such as reduction of inventory in the IIoT factory, or the identification of an opportunity that can help the IIoT organization move ahead of the competition such as the desire of increasing share-of-wallet in the sales

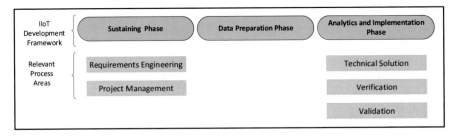

Fig. 7.1 IIoT analytics development Framework

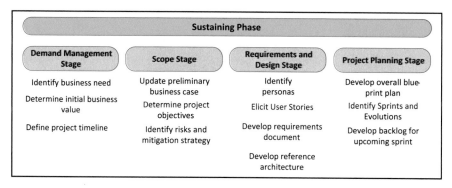

Fig. 7.2 Sustaining and underpinning phase

organization by using advanced analytics. As the business opportunity is identified, the analysts need to understand if the opportunity is a data-intensive activity, i. e. if analysis of historical data associated with the opportunity can solve the issue. This means, that a high-level assessment of the type of data associated with the opportunity, data availability, and data quality need to be conducted. Once the business problem or opportunity and also the data availability have been assessed, it is required that the business value is defined for the opportunity which typically is either revenue generated or cost cutting or savings. During the demand management stage, it is also important to define an initial timeline for the analytics project and an initial estimate of size, effort, and cost.

During the *Scope stage* the business case for the analytics project is updated and refined by developing a deeper understanding of the economic benefits of the project to the IIoT organization. A more detailed definition of the analytic project objectives is prepared, and a formal definition of the potential risks associated with the project is created.

During the *Requirements and Design stage*, the users of the analytic system in the IIoT organization are identified and are called *Personas* and they represent the typical roles that are expected to use the system. Once the personas are identified, the User Stories are created. User stories are high level functional capabilities and quality attributes that are expected from the analytic system. Each person will have a set of user stories that are relevant to him/her. Once the User Stories are defined these are decomposed into *Requirements* that are defined at a lower level granularity. Finally, a reference architecture for the analytic system is created depending on the requirements to be implemented. As discussed in Chap. 1, the IIoT analytics architecture will serve as a basis for the implementation of the analytic system, but a specific reference architecture based on this analytics architecture is defined.

During the *Project Planning stage,* the overall project plan for the analytics project is defined. The plan consists of a high-level view of the entire project. Nevertheless, the best way to develop an analytics project is to use an Agile Development lifecycle to ensure continuous customer/user involvement, to ensure rapid re-alignment on potential changes in the requirements, and to show continuous progress in the project. In Agile software development, detailed plans are prepared for the upcoming sprint cycle.

Requirements Engineering

As mentioned above, the Requirements Engineering process can be sub-divided into two process areas: Requirements Development and Requirements Management.

The purpose of *Requirements Development* is to elicit, analyze, and establish customer, product, and product component requirements. The purpose of Requirements Management is to manage requirements of the project's products and product components and to ensure alignment between those requirements and the project's plans and work products.

The Requirements Development (RD) process area is fundamental in the development of any software-intensive system and it is no exception for the development of advanced analytic applications. The IIoT organization needs to set in place a robust RD process that addresses the following items [3, 12].

- Elicit stakeholder needs, expectations, constraints, and interfaces for all phases of the lifecycle.
- Transform stakeholder needs, expectations, constraints, into prioritized customer requirements.
- Establish and maintain product and product component requirements.
- Identify interface requirements.
- Establish and maintain scenarios and user stories.
- Establish and maintain a definition of required functionality and quality attributes.
- Analyze requirements to ensure that they are necessary and sufficient.
- Analyze requirements to balance stakeholder needs and constraints.
- Validate requirements to ensure the resulting product will perform as intended in the end user's environment.

From the points above mentioned the elicitation and maintenance of different levels of requirements becomes evident. First, it is important to elicit the requirements for the analytic application, and then these requirements are further decomposed to a level of granularity that allows the analytic developers to develop them and eventually the testers to test these. Figure 7.3 shows a proposed requirements decomposition in a tree hierarchy that can be used when collecting and documenting requirements.

The requirements tree shown in Fig. 7.3 is a representation of the sample requirements for an analytic application for an IIoT organization that wishes to develop a recommender system for their Service Engineers to advise their installed base customers on potential services and spare parts that may be of interest to them. Details on this application are given in Chap. 6. The requirements tree shown in Fig. 7.2 follow the principles of a formal Requirements Development process and is an instantiation that the author has successfully used in the development of analytic

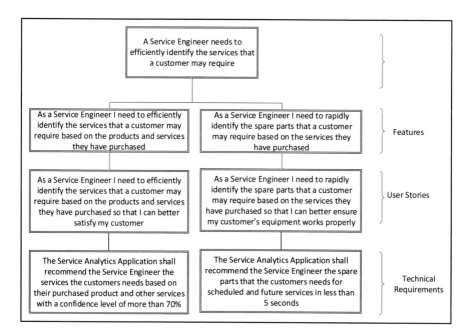

Fig. 7.3 Requirements decomposition for analytics projects

and software applications. It is important to notice that different authors and professionals may have a variation of this tree structure and the names of each tree level may also be slightly different. For example, some authors may refer to the "User Stories" as "Use Cases" and so on.

The tree structure of requirements is generated through interviews with the Service Engineers (personas of interest or potential users of the system) that will be the target of the service analytics application. This can be done through a design workshop that typically lasts 2 to 3 days. Notice the way each requirement level has been defined, as the language or format used are important to keep neutrality, improve understanding of the stakeholders, and increase precision for development and testing of the requirements. First, the *Needs* are determined for the system users and they represent the core functionality that aims at satisfying the expected business outcomes. Each need is decomposed into *Features* that describe in more detail the need adding a quality attribute to the need such as "efficiently" or "rapidly". At the third level, the more detailed *User Stories* are defined where in addition to the quality attributes, the reason behind the user story is provided. Once again, notice the language used. Finally, the *Technical Requirements* are stated at the bottom of the tree and they show explicitly the requirement with its fit criterion (way to measure the requirement) [12]. Notice that the quality attribute in the technical requirement is quantified so that the developers and testers have an explicit way to develop and test the requirement, as a benchmark.

The *Requirements Management* (REQM) process area is needed to manage requirements of the project's products and product components and ensure alignment between those requirements and the project's plans and related work products. Some of the primary activities established by a robust REQM process include:

- Develop an understanding with the requirements providers on the meaning of the requirements.
- Obtain commitment to requirements from all project participants.
- Manage changes to requirements as they evolve during the project.
- Maintain bidirectional traceability among requirements and all work products.
- Ensure that project plans and work products remain aligned with requirements.

The Requirements Management process area is fundamental in the implementation of an Agile development lifecycle. This is true because as changes are incorporated into the development of the analytics project during the sprint meetings with the customer, a strong Requirements management process is needed to keep track of the changes in the requirements, additions, or modifications. These changes need to be continuously monitored and reflected into the project plan and backlog dashboards that are used to monitor the project.

Project Management *Process*

The purpose of the *Project Management* process area is to establish and maintain plans that define project activities. As presented above, the Project Management process is sub-divided and Project Planning and Project Monitoring and Control.

Project Planning is a process area that aims at developing and maintain plans that help carry out the project activities. Analytic projects for the IIoT organization need to show quick results and therefor it is recommended that agile development lifecycles are used such as Scrum [5] or Kanban [2]. An agile development approach such as Scrum converts the milestones identified in the project into short sprints (typically one to 4 weeks in length), producing customer-ready product. Kanban is built for smooth and continuous delivery of customer value. Kanban controls the flow and quality of work to discover and resolve issues immediately. Kanban limits the work in progress so that churn doesn't build up and the team and product can adjust to market shifts daily [2]. Estimation of the size, effort, time, and cost of an analytics project is one important element in the project planning activity. Estimation is done at the beginning of the project, and it is referred to as "initial estimation", and then is continuously done as the project evolves, especially if an agile development lifecycle is used [6]. A defined work breakdown structure or product backlog represents the activities necessary to develop the analytics project. Good practices of estimation require the use of reliable historical data. If reliable historical data are not available, using an Agile development approach can be helpful to collect data at each development cycle or sprint and use it in subsequent sprints. A useful Agile approach to estimation that can be used to estimate analytic projects is the Planning Poker method [5]. This approach can be used regardless if the IIoT organization has reliable historical estimation and project management data or does not have them. Developing a project plan involves determining the budget and schedule of the project using the results from the estimation. It is essential to identify the project risks in an IIoT project, the impact of the risks, the probability, the overall exposure (probability * impact), and develop a risk mitigation strategy. The project plan includes identifying the resources for the project, the needed knowledge and skills, and planning for the stakeholders' involvement [3].

The purpose of the *Project Monitoring and Control* process area is to provide an understanding of the project's progress so that appropriate corrective actions can be taken when the project's performance deviates significantly from the plan [3]. Once the plan for the IIoT analytics project has been established and commitments from all project stakeholders have been made, it is important to monitor how the project moves forward. The Project manager needs to monitor all project planning parameters (that are the indicators of the project's progress) against the project plan, monitor all commitments from stakeholders, monitor risks, monitor project stakeholders' involvement in the project, conduct necessary progress and milestone reviews (scrum meetings and sprint review meetings). Another important element during the monitoring of the project is to collect and analyze the issues that emerge from an IIoT analytics project and determine the corrective actions needed to address the issues. Once the corrective actions have been taken, they need to be managed to closure. The project monitoring and control process area is also fundamental when using an agile development lifecycle model, as it needs to reflect the continuous changes that occur at the end of each sprint in the development project.

Data Preparation Phase

The second phase in Fig. 7.1 include the Data Preparation activities that are required to ensure the data required for the analytics are. In the data preparation activity, the identification of the required data sources is conducted, the data modeling activity is conducted, the data are then extracted into a Data Lake/Data Warehouse, and then the data is cleaned and preprocessed to make it ready for the analytics activity.

During the Data Preparation phase the data available to address the pain point or opportunity are identified. It is important to identify the sources of data that can be used. Once these data have been identified, it is important to understand the data attributes in the source data repositories that are relevant. First, the relevant data sources to address the opportunity are identified. The data attributes are examined with domain experts in the field of the analytics application to understand the data. A data dictionary is prepared that defines each table and data attribute. The data profiling activity is conducted to understand basic statistics such as highest value, lowest value, mean, median, number of missing values, histograms, etc. A data model is then built based that is needed to extract the data that will be used for the analytics application. As shown in Chap. 1, the data is then extracted from its original sources so that can be ingested into the Data Lake/Data Warehouse of the

IIoT analytics system shown in Fig. 1.3. After the data is ingested then it is placed in the landing area or staging area of the Data Lake/Data Warehouse. Once the raw data is in the staging area, then the data is transformed and preprocessed so it is ready to be used by the analytics application. An important element is the data refreshing activity where new data generated and stored in the source repositories needs to be uploaded into the Data Lake/Data Warehouse at pre-determined intervals of time, as needed by the application. The data preparation phase is shown in Fig. 7.4.

Large, complex, multi-national organizations face a major challenge defining the data models that they will use in their IIoT analytic applications. This difficulty raises from the fact that many geographically dispersed units often have differences on how they define the data attributes in their local data repositories [11]. For example, in the same organization, two sister companies may be monitoring a value such as "humidity". One company may be calling this attribute humidity level, while the sister organization may be calling it "moisture" level. And many examples like this can be found. Figure 7.5 shows a high-level perspective of the layering of data in an IIoT enterprise. The IIoT enterprise consists of many units that include factories and other business located in different geographic regions. Some of the businesses have either a geographic affinity or they may build similar products or collaborate to build parts for the same products. It is easier for similar locations to define data attributes in similar fashion and sometimes they even share enterprise resource planning and other systems, where data tables and attributes are defined similarly.

Starting from the left-hand side of Fig. 7.5, a specific business or factory has a set of source data repositories used to manage their businesses and make decisions. The data is stored in their local data repositories in structured tables with their corresponding data attributes. In the IIoT enterprise, the data from the businesses is ingested "as-is" into the Data Warehouse or Data Lake into the "Staging" area. As data is stored into the staging area, then it is transferred into expanded table into the "Consolidation" area. The data consolidation is a process that merges all of that data

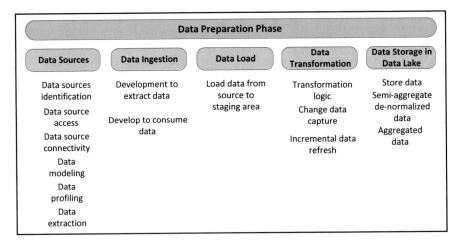

Fig. 7.4 Data preparation phase

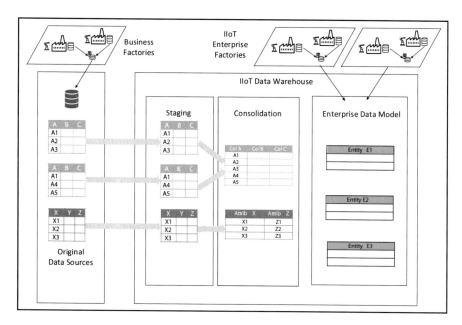

Fig. 7.5 Data layering in an IIoT enterprise

in the staging area, removes any redundancies, and cleans up any errors before it gets stored in the consolidation area. This process allows to make data storage more efficient, increases data consistency, and improves data quality. As different businesses, especially distributed businesses in the IIoT enterprise, place their datasets into the consolidation area of the data warehouse, it is important to bring the data into one data semantic umbrella and ensure the data attributes are defined in a consistent manner so that the data can be used for enterprise analytics. This data layering, pre-processing, consolidation, and homogenization can be time consuming and require a great deal of effort and communication among the IIoT enterprise stakeholders, but it is essential to be able to have high quality and trustworthy analytics in the enterprise.

Analytics and Implementation Phase

The analytic and implementation phase refers to those activities that analytics developers follow to deliver the best analytics solution(s) to solve the business problem. From the Software Engineering perspective, three process areas are relevant. Technical Solution, Verification, and Validation. The purpose of Technical Solution is to select, design, and implement solutions to requirements. Solutions, designs, and implementations encompass products, product components, and product related lifecycle processes either singly or in combination as appropriate. The purpose of

Verification is to ensure that selected work products meet their specified requirements. The purpose of Validation is to demonstrate that a product or product component fulfills its intended use when placed in its intended environment.

During the analytics and implementation phase of the analytics project, the IIoT organization focuses on how to address the business pain point, problem, or opportunity. This activity is primarily conducted by the data scientists, data development developers, and visualization and user interface developers. Figure 7.6 shows the stages in the analytics and implementation phase.

During the IIoT analytics stage, there are four possible analytics approaches that models can address: (a) Descriptive Analytics; (b) Diagnostic Analytics; (c) Predictive Analytics; (d) Prescriptive Analytics, as discussed in Chap. 2. *Descriptive analytics* refers to approaches that analyze historical data using Statistical methods that summarize historical data and convert it into a form that can be easily understood by domain experts. *Diagnostic analytics* aid domain experts to explore deeper into an issue at hand so that they can arrive at the source of a problem and answer the question of why something happened. *Predictive analytics* helps businesses to forecast trends based on the current events. *Prescriptive analytics* prescribes domain experts what action to take to eliminate a future problem or take full advantage of a promising trend.

Cognitive Analytics is another type of analytics that is outside the scope of this book. *Cognitive Analytics* attempts to mimic the human brain by drawing inferences from existing data and patterns, extracting conclusions based on existing knowledge bases and then inserting this knowledge back into the knowledge base for future inferences. This then becomes a self-learning feedback loop [1].

An essential part of the analytics project is the visualization element. During this stage, the domain experts that will use the system, the data scientists, and the visualization experts work together to develop wire frames (that are mock up visuals) the dashboards, visualizations, and graphic user interfaces that the domain experts will use to interact with the system and will help them make their decisions. As the

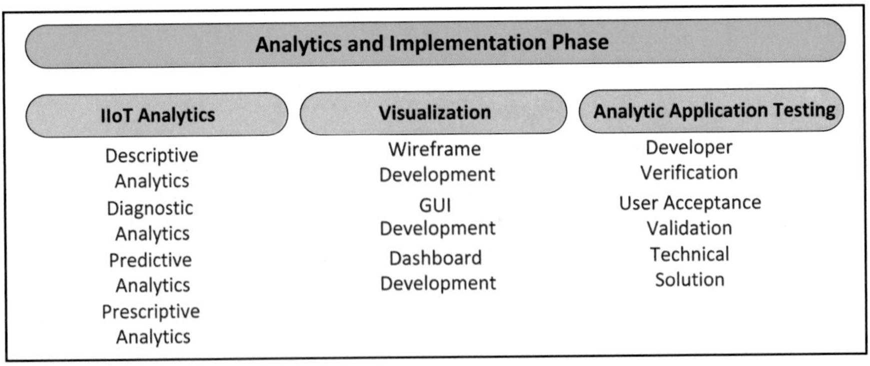

Analytics and Implementation Phase

IIoT Analytics	Visualization	Analytic Application Testing
Descriptive Analytics	Wireframe Development	Developer Verification
Diagnostic Analytics	GUI Development	User Acceptance Validation
Predictive Analytics	Dashboard Development	Technical Solution
Prescriptive Analytics		

Fig. 7.6 Analytics and implementation phase

project development evolves, so the visualizations and dashboards to reach their final stage.

Technical Solution Process

The purpose of the Technical Solution process area is to design, develop, and implement solutions to the identified requirements for the IIoT analytics project. This process area focuses on identifying, evaluating, and selecting solutions or design approaches that satisfy the identified requirements in the project. The TS process area also focuses on developing detailed designs for the selected solutions and implementing the selected designs. By having a robust technical solution process, the IIoT organization can evaluate the best design approach to systematically address the analytics problem or opportunity [3].

Verification and Validation Processes

Once the system is completed, the verification stage takes place where two primary activities occur as shown in Fig. 7.4. Verification of the analytic solution refers to ensuring the analytic system and its components work according to the specification. The verification is conducted by an independent testing team. Validation on the other hand ensures that the analytic solution will fulfill its intended use. A successful validation ensures that the product and its components will work as intended in the deployment environment when used by the users [3].

The Verification process area involves the following: verification preparation, verification performance, and identification of corrective actions to found defects. Verification includes verification of the product and intermediate work products against all selected requirements. Verification is inherently an incremental process because it occurs throughout the development of the product and work products, as they evolve during the agile development lifecycle. Verification starts with verification of the requirements, progressing through the verification of the evolving work products, and culminating in the verification of the completed product.

Testing Machine Learning models for advanced analytic applications for IIoT enterprise refers to testing the performance of machine learning models in terms of accuracy, confidence, and precision of the results of the model. The principles of the verification process applied to software development are also applicable to Machine Learning models. Machine Learning models need to be tested as conventional software development from the Quality Assurance perspective. Techniques such as black box and white box testing also apply to Machine Learning models. Black box testing refers to testing a software module based on the expected functionality of it rather than the internal structure and code. When applied to Machine Learning models, black box testing means testing the models without knowing the internal details such as features of the Machine Learning model, the algorithm used to create the model, or how was programmed. "The challenge, however, is to identify the test oracle which could verify the test outcome against the expected values known beforehand" [9].

In conventional software development, a frequently invoked assumption is the presence of a test prediction which is a set of testing mechanisms including the testing program which could verify the output of the software module against the expected value which is known beforehand. In the case of Machine Learning models, there are no expected values beforehand as the Machine Learning modules' output is usually a prediction. Given that the outcome of Machine Learning models is a prediction, it is not easy to compare or verify the prediction against an expected value which is not known beforehand. Nevertheless, during the model building phase, data scientists test the model performance by comparing the model predicted outputs with the actual values of known data points (after the model has been trained with a large subset of historical data then it is tested with a smaller subset of that historical data). This is not the same as verifying the Machine Learning model for any input where the expected value is not known [9].

In such cases where test oracles are available the concept of pseudo oracles is introduced. Pseudo oracles represent the scenario where the outputs for a given set of inputs is compared with each other and the correctness can be determined. Consider the following scenario. A program for solving a problem is coded using two different implementations with one of them to be chosen as the main program. The input is passed through both implementations. If the resulting output is the same or comparable (falling within a given range), the main program could be said to be working as expected or correct program. This is the way quality assurance or verifications can be performed with Machine Learning models, where two similar Machine Learning models outputs can be compared given a certain set of historical data points. Kumar [9] discusses approaches that can be used to perform back box tests of machine Learning models.

First, testing **Model Performance** refers to testing two similar Machine Learning models with the test data set or new data set and compare the model's performance in terms of parameters such as accuracy, confidence, recall, or precision. This is the most used black box verification technique for Machine Learning models.

Second, in the **Metamorphic Testing** approach, one or more properties in the model are identified that represent the metamorphic relationship between input-output pairs. For example, if a Machine Learning model is built and predicts that ML model is built and predicts that the likelihood of a machine exhibiting a fault is based on different predictor variables such as age, frequency of maintenance, usage, etc. Based on the detailed analysis, it is derived that given the machine has all maintenance activities met during the year, and its usage is high, the likelihood of the machine exhibits the fault increases by 5% with an increase in its age by 3 years. This could be used to perform metamorphic testing as the property, age, represents the metamorphic relationship between inputs and outputs. The test cases for the Machine Learning model could be:

(a) Given the machine is properly maintained and its usage is high, determine the likelihood the machine will exhibit the fault given that the machine is 10 years old.
(b) In the input then, we can increase the age by 5 years. Then we would expect the probability of exhibiting the fault should increase by more than 5%.
(c) Increasing the age of the machine by 10 years, the likelihood that the machine exhibits the fault should increase by more than 15% but less than 20%.

Third, in the **Dual Coding** approach, the idea is to build different models based on different algorithms and comparing the predictions from each of these models given a particular input data set.

The Verification and Validation process areas are similar, but they address different issues. Validation demonstrates that the product will fulfill its intended use, whereas verification addresses whether the work product properly reflects the specified requirements. In other words, verification ensures that the product is built correctly while validation ensures the right product has been built.

Validation activities can be applied to all aspects of the product in any of its intended environments. The methods employed to accomplish validation can be applied to work products as well as to the product and product components. The work products (e.g., requirements, designs, and prototypes) should be selected on the basis of which are the best predictors of how well the product and product component will satisfy user needs and thus validation should be performed early and incrementally throughout the product lifecycle. The validation environment should represent the intended environment for the product and product components as well as represent the intended environment suitable for validation activities with work products. Validation demonstrates that the product, as provided, will fulfill its intended use. Validation activities use approaches similar to verification (e.g., test, analysis, inspection, demonstration, or simulation). Often, the end users and other relevant stakeholders are involved in the validation activities. Both validation and verification activities often run concurrently and may use portions of the same environment.

Agile Kanban Development Lifecycle

Agile Kanban is a methodology that allows to develop high quality software solutions that show value to customers quickly within the expected budget. The Agile Kanban approach involves developers and potential users early in the development process and brings them along the development with the so called "short "sprint meetings" to initially develop a "minimum viable product" (MVP) and then build on it to continue enhancing its functionality. Defining the MVP helps narrow down a target development and testing activities and allows to develop a feasible deadline

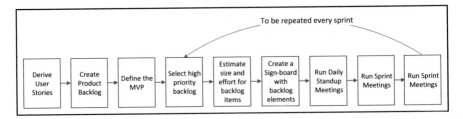

Fig. 7.7 Overview of steps in Agile Kanban method

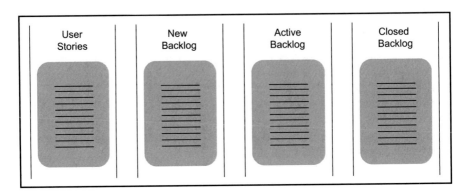

Fig. 7.8 Agile Kanban signboard

to complete the MVP, which is a product that can be released. The work needed to achieve the MVP is then essential and will visibly contribute to an initial product to be released to the customer. Figure 7.7 shows a simplified overview of the Agile Kanban methodology.

A product backlog is a list of features derived from the user stories, changes to existing features, defect fixes, infrastructure changes or other activities that a team may deliver in order to achieve a specific outcome. The initial outcome is the MVP and as the project makes progress the backlog is the set of activities identified to continuously improve the MVP towards the final deliverable.

An easy way to define the MVP is running what is called "bucketing affinity" exercise that helps sort out the backlog entries into four priority buckets: (a) the must have; (b) should have; (c) like to have; (d) nice to have. Once this sorting is completed the selection of the highest priority backlog items constitute the task required to develop the MVP. Once this is done, it is clear what elements of the backlog or tasks the team needs to focus to deliver the MVP and the future backlog.

At this point the team then will utilize an estimation method such as the Planning Poker, or Monte Carlo to estimate the size and effort of each element of the backlog. The initial estimation for each element of the backlog is to complete the MVP.

With the backlog items defined, a signboard is prepared that has four main columns: (a) user story column; (b) the new backlog column; (c) the active backlog column; (d) the closed backlog column. Figure 7.8 shows the layout of the signboard, which can be implemented physically or virtually. As the development work makes progress, during the daily update (or also known as SCRUM) meetings the team reviews the daily backlog and makes appropriate updates on progress. In the user story column, the set of user stories to be developed are placed and under each user story the backlog items associated with the user story are placed in the new backlog column. Backlog items that are currently being worked on in the sprint are placed in the active backlog column. When a backlog item is completed then the item is closed and placed in the closed column.

Barriers for the Implementation of Analytic Projects in the IIoT

Analytic projects that are developed to assist decision makers in the IIoT enterprise need to be carefully conceived and carried out to be successful. There are several potential barriers that can prevent these analytic projects to be successfully implemented and used by the IIoT knowledge workers and these need to be carefully addressed throughout the development lifecycle and after they are initially implemented in the organization and they are listed below.

Lack of Clear Business Value

As mentioned earlier in chapter and previous chapters it is of paramount importance to define a clear and concise business case that supports the development of an advanced analytics project that fuels decision making in an IIoT enterprise. It is also important to perform a cost/benefit analysis for the project. Costs need to be considered such as development costs, costs of involving domain experts, implementation costs, and maintenance and support costs of final solution. Similarly, expected benefits of the analytics solution need to be reflected such as cost savings, or increase in hit rates, or increase in service sales, or increase in product sales, reduction of processing times, and others. Lack of a clear and positive business value for the project that develops the analytics solution will decrease senior management support for the project and will cause it to fail. Moreover, if the users do not have a clear sense of benefit of the application, they may start using the application but in time they will reduce its utilization until the application is not used at all.

Absence of Large User Base

The objectives of the initial Design Workshops at the start of the analytics development project are to define the personas or future system users, their needs, their pain points, their opportunities, and the functionality that needs to be developed in terms of User Stories. Each persona or SME will have associated their own user stories that reflect the way they will be interacting with the analytics system. So, it is important to identify not only the specific personas, which identify the generic roles of the users of the system. It is also necessary to define the number of potential users under each persona definition so that the total expected number of users can be estimated. Although there is no "magic" number of users that will make an analytics solution succeed, a sizable user base will create and maintain the momentum behind the analytic solution. Moreover, if there is potential to hire new users due to the expanded business capabilities created by the analytics solution this is an extra element to keep the momentum on the use of the system.

Takes Too Long to Develop the Solution

The development lifecycle of software applications in the software industry continues to shorten. It is no exception for the development of analytic solutions, dashboards, and visualizations that are used in an IIoT enterprise. The expectation from potential users in the IIoT enterprise is that development of solutions should be completed very fast, in short development cycles, and deliver the so called "minimum viable product" (MVP) [2] and then build on top of it to enhance the functionality of the solution. "Typically, MVP items are basic functionality that customers expect before they'd even try your product, plus just enough differentiation to determine whether your new release is desirable" [3, pp. 27]. The total delivery time of an analytic solution includes the time it takes to clearly define the opportunity or pain point to address, the time to develop a solid business case, the time to define the personals and user stories, and the time to develop more detailed requirements. It is important to move fast throughout the scoping phase and then the development phase. This can be achieved by using an agile methodology and creating sprint review meetings keeping the users/customer engaged at all time and continuously delivering results, preferably every two weeks or so. Moving fast will maintain the solution stakeholders fully engaged in the project.

Organization Focused on Short-term Gains

For the most part analytic development projects, fully realize their benefits in a midterm time horizon. This means that the full development, user acceptance testing, and implementation of an analytics solution easily takes 8 to 12 months. Once deployed, the full economic benefits can be realized after six months to a year of using the solution. For this reason, the IIoT organization needs to be focused on a medium-term gains paradigm and give six months to a year time horizon to begin seeing the economic benefits of the analytic solution developed and implemented. An important aspect of measuring the economic benefits is that a benchmark or "stake in the ground" needs to be defined to them measure the state of the process before the implementation of the analytic solution and then after the analytic solution has been implemented.

High-level Complexity

High level complexity can become an obstacle for the development of an advanced analytics solution in the IIoT enterprise. High-level complexity can manifest in the definition of the business case of the solution. For example, it is clearer to explain the benefit of an analytics solution when this benefit is associated with revenue generation rather than cost savings. Examples of revenue generating analytics solutions include: (a) an analytics system that allows new sales of products; (b) a recommender system that generates new service opportunities; (c) a recommender system that helps expedite account receivable issues; (d) an analytic system that helps opening new markets. Although economic benefits of analytic solutions associated with cost savings are very valuable, it is more difficult to explain in a straightforward manner the achieved benefit. For example, an analytic solution for inventory analytics that reduces the costs of inventory in a manufacturing operation needs a lot more explanation to senior management and often needs to be paired with a secondary benefit such as on-time-delivery (OTD) of products that can be associated with some kind of revenue generation.

Another element of complexity that needs to be managed and can result in not trusting the system is how the analytic solution is derived. Using complex analytic algorithms and analytic methods that are difficult to explain and visualize, increase the skepticism of domain experts in the proposed solution. For example, it is a lot easier to explain the results of an analytic solution derived from a decision tree or random forest algorithm rather than the result from a neural network algorithm. Subject matter experts using the analytic solution need to have a good level of confidence in the solutions proposed by the system.

Organizational Readiness for Change

The implementation of a new analytic solution in any organization carries out an important element of disruption, even in an IIoT enterprise that may be more open to change as employees are frequently exposed to new technologies. The implementation of new analytic solutions results in changes in the process they are integrated.

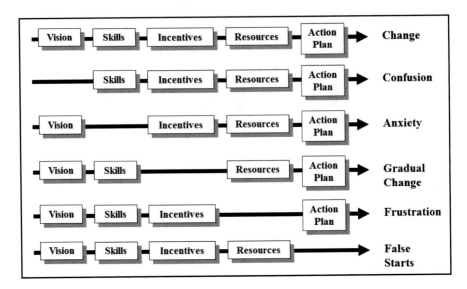

Fig. 7.9 Organizational change matrix as per Debou [7]

This means that the users and adjacent professionals need to adapt to new processes and a new way of interacting with each other. For this reason, the organization needs to be open to embrace change. When an organization is used to carry out their productive activities in a certain way, members of the organization have difficulty to make changes in the way they work [10]. Change occurs when any part of the organizational system is modified or replaced. Change means replacing what is established in favor of something new. In a software process improvement activity, old and established development practices are replaced by improved, streamlined, and more efficient practices and processes. Even if the new practices and processes selected for adoption enhance the organization's operations, there is always a tendency for the organization to resist the change. Organizational change readiness refers to the capacity that an organization possesses to respond to new challenges in its operational environment.

Organizational change readiness refers to the capability that an organization exhibits at a point in time to adopt a new behavior and respond to new challenges in its operational environment. The Software Engineering Institute (SEI) and Debou [7] have defined a change matrix that identifies the necessary elements for change and how the lack of these elements results in frustrated and ineffective efforts, as shown in Fig. 7.9.

Necessary elements for change, and hence for the implementation of a successful Software Process Improvement (SPI) program include: (a) *vision* on where we want the organization to be with the proposed SPI program; (b) *skills* in the organization necessary to achieve the improvements; (c) *incentives* and *dis-incentives* provided by senior management to the organization to change its behavior; (d) *resources* needed to successfully conduct the SPI activity; (e) an *action plan* that guides the

SPI program activities. Figure 7.8 shows that the lack of *vision* in an SPI program causes confusion in the organization. The lack of *skills* necessary to conduct the SPI activities causes anxiety in the organization. The lack of *incentives* or *dis-incentives* to members of the organization affected by the SPI activities cause a delay in the SPI program. The lack of adequate *resources* (tools, time, people, etc.) causes great frustration in an organization trying to implement a successful SPI program. And finally, the lack of an *action plan* causes false starts and disorientation in the SPI program. When an organization is faced with a severe change prospect such as the introduction of a new technology people in the organization may suffer from similar reactions as the stages of loss and grief which include denial and isolation, anger, depression, and acceptance [8].

References

1. Bharadwaj, V. (2016). https://www.quora.com/What-is-cognitive-analytics
2. Brechner, E. (2015). *Agile project management with Kanban*. Washington, DC: Microsoft Press, Redmond.
3. Bourque, P. & Fairley, R. E., (Eds.). (2014). Guide to the software engineering body of knowledge. Version 3.0, *IEEE Computer Society*. www.swebok.org.
4. Chrissis, M. B., Konrad, M., & Shurm, S. (2011). *CMMI for development*. Addison-Wesley, Third Edition.
5. Cohn, M. (2006). *Agile estimating and planning*. Upper Saddle River: Pearson Education Inc..
6. Dagnino, A. (2013, May 18–26). Estimating Software-Intensive Projects in the Absence of Historical Data. In *35th International Conference on Software Engineering (ICSE 2013)*, San Francisco, CA, USA.
7. Debou, C. (2009). *Managing change: The human factor in process improvement initiatives*. Stuttgart: Presentation from Kugler Maag CIE.
8. Kubler-Ross, D. (1969). *On death and dying*. Simon & Schuster (Eds.), USA.
9. Kumar, A. (2018). *QA: "Black box testing for machine learning models"*. The AI Zone. https://dzone.com/articles/qa-blackbox-testing-for-machine-learning-models
10. Massey, A. P., Montoya-Weiss, M. M., & Brown, S. A. (1988). Managing technological change when change is mandatory. *IEEE International Conference on System, Man, and Cybernetics, 5*, 4758–4762.
11. Pomp, A., Paulus, A., Kirmse, A., Kraus, V., & Maisen, T. (2018, September). Applying semantics to reduce the time to analytics within complex heterogeneous infrastructures. *Technologies, 6*(86), 1–29.
12. Robertson, S. & Robertson, J. (1999). *Mastering the requirements process*. Addison-Wesley.
13. Shruti, S. & Hasteer, N. (2016, April 29 and 30). A comprehensive study on state of scrum development. In *International Conference on Computing, Communication and Automation Conference, IEEE* (pp. 867–872). Noida, India.
14. Stobierski, T. (2018, June 8). *Data analysis and project management: How analysts can benefit from project management techniques*. North Eastern University, Graduate Programs. https://www.northeastern.edu/graduate/blog/data-analyst-and-project-management/

Conclusions

The development of new computing, communication and information technologies, Cloud Computing, data security, almost unlimited data storage capabilities, wireless and sensing technologies, high-performing and sophisticated analytic algorithms, and sophisticated user interfaces have contributed to the development of the Internet of Things and the Industrial Internet of Things. The Internet of Things is a proliferation of smart objects or products that connect the physical world to the digital world in such a way that humans remotely interact with these devices to enhance their consumer experience, dialogue with these devices, monitor their performance, and remotely guide these devices to perform specific tasks. The Industrial Internet of Things is the application of the Internet of Things principles into vastly larger industrial world such as manufacturing, power generation, transmission, and distribution, supply chain management, anomaly detection of industrial systems, and many others to increase productivity, enhance services to customers, and in general make the industrial world more efficient, more environmentally friendly, more sustainable, and more human friendly. The IIoT enhances systems availability, intelligence and connectivity within a world that needs to manage very high volumes of data and requires continuously scaling its resources such as in mining operations, power generation, the smart city, and others. It is important to notice that the IIoT operates in a world where failures could be very costly. IIoT can massively improve connectivity, efficiencies, scalability, time savings, and cost savings in industrial organizations.

This book presents industrial real-world examples of IIoT applications where advanced analytics and machine learning models from AI have been used to address a business opportunity. In addition to the examples presented many more examples are available that have not been discussed in this book. For example, as this book was being completed the author was part of a team that was developing an analytic solution for an IIoT organization to track how the COVID-19 pandemic was globally affecting businesses in the organization. The objective of this project is to help the organization to develop a global and local strategies to reduce the impact of the virus to employees, contractors, and other stakeholders, and also adapt to changes

© Springer Nature Switzerland AG 2021
A. Dagnino, *Data Analytics in the Era of the Industrial Internet of Things*,
https://doi.org/10.1007/978-3-030-63139-0

in supply chain, manufacturing, delivery, and other production activities in the organization.

As the topic of the IIoT was explored, there are some prominent concepts that emerge in this book. It is my opinion that industries that do not adopt the IIoT concepts and make them part of their business models and operations will slowly become obsolete as they will lose customers and competitors that will adopt IIoT will take over. Industries need to understand how the IIoT can change their business models and make them increasingly competitive. New business models such as selling capabilities instead of products will become the norm. Selling services that will liberate customers from managing their assets, have a maintenance department, and buy traditional equipment will become the norm. IIoT organizations will be able to provide productivity asset equipment to their customers that changes capabilities and adapts to production environments virtually through the Cloud in a secure manner.

It is important that industries that decide to follow the IIoT path adopt a continuous improvement modus operandi so that their management and employees thrive in a change environment that spans from their business models, to their processes, the products and services they provide, the technology required to have an adaptable, robust, and sustainable IIoT platform. These organizations need to implement a strong change management mindset to be open to adapt to the changing competition and quickly adjust to customers' demands and expectations. An element of change that IIoT organizations need to keep particularly in focus is to excel at becoming more and more environmentally friendly and safety driven using the IIoT technology and processes.

Although IIoT brings a high level of technology and automation, the IIoT enterprises need to understand that their employees or human capital are the most important asset. The knowledge that employees gain in the IIoT enterprise needs to be recognized, expanded, and leveraged by IIoT solutions. As these solutions use sophisticated computer, communication, and sensing technologies fueled by powerful analytics and artificially intelligent algorithms, these must be developed keeping in mind that they serve the employees. The employee, subject matter expert, and domain specialist in the IIoT organization should use the IIoT solutions to make decisions and to enhance job satisfaction. It is also my believe that if used properly, the IIoT will generate very satisfying jobs and increase employment.

An essential element in the IIoT are data. This book does not explore in depth all data aspects that need to be considered to have healthy analytics. Data is the raw material that analytic models and AI uses to identify patters, predict, project, forecast, prescribe and learn. For these reasons it is important to pay special attention to data integrity and use "good data" to feed the analytic models. Having high quality algorithms and having poor quality data results in poor quality results and analytics. Data can be incomplete, inconsistent, not collected or registered correctly, can be biased, values out of range, textual data not properly recorded, etc. For this reason, it is essential that the IIoT puts in place processes and strategies to ensure high data quality. Define robust processes for data collection and ingestion into data lakes or data warehouses is essential. Creating consistent and reusable data models is also very important.

The IIoT enterprise needs to have businesspeople, domain experts, data scientists, and visualization experts working together. The businesspeople will be able to define high business value opportunities and business models that can be crystalized by the IIoT. The domain experts have the knowledge about products, services, customers, and processes that are essential to create analytic models and processes to fulfill the business opportunities identified by the businesspeople. The data scientists have the knowledge in statistics, machine learning and artificial intelligence and will work with the domain experts to create the IIoT analytic solutions that fuel the IIoT systems. Finally, the visualization and user experience (UX) experts, will be able to bridge the gap between the analytics world and the users and domain experts by providing user friendly dashboards and visualizations that translate the results obtained by the analytic models.

In summary, potential of the IIoT technologies is monumental and we can only imagine what will happen in the next decades. For this reason, managers of industrial organizations need to monitor how IIoT is reinventing the industry, especially as Big Data Warehouses become more ubiquitous and Analytics and Artificial Intelligence modeling fuel the IIoT enterprise

Index

© Springer Nature Switzerland AG 2021
A. Dagnino, *Data Analytics in the Era of the Industrial Internet of Things*,
https://doi.org/10.1007/978-3-030-63139-0

Printed in the United States
by Baker & Taylor Publisher Services